MATHEMATICAL THEORY IN PERIODIC PLANE ELASTICITY

Asian Mathematics Series

A series edited by Chung-Chun Yang, *Department of Mathematics, The Hong Kong University of Science and Technology, Hong Kong*

Volume 1

Dynamics of Transcendental Functions

Xin-Hou Hua and Chung-Chun Yang

Volume 2

Approximate Methods and Numerical Analysis for Elliptic Complex Equations

Guo Chun Wen

Volume 3

Introduction to Statistical Methods in Modern Genetics

Mark C.K. Yang

Volume 4

Mathematical Theory in Periodic Plane Elasticity

Hai-Tao Cai and Jian-Ke Lu

MATHEMATICAL THEORY IN PERIODIC PLANE ELASTICITY

Hai-Tao Cai

Department of Mathematics
Central South University of Technology, Hunan, P.R. China

and

Jian-Ke Lu

Department of Mathematics
Wuhan University, Wuhan, P.R. China

GORDON AND BREACH SCIENCE PUBLISHERS
Australia • Canada • France • Germany • India • Japan
Luxembourg • Malaysia • The Netherlands
Russia • Singapore • Switzerland

Amsteldijk 166
1st Floor
1079 LH Amsterdam
The Netherlands

British Library Cataloguing in Publication Data

Cai, Hai-Tao
Mathematical theory in periodic plane elasticity. – (Asian mathematics series ; v. 4)
1. Elasticity – Mathematics 2. Periodic functions
3. Differential equations
I. Title II. Lu, Jian-Ke
531.3′82′0151535

ISBN 13: 978-9-0569924-22

Contents

Introduction to the Series

The *Asian Mathematics Series* provides a forum to promote and reflect timely mathematical research and development from the Asian region, and to provide suitable and pertinent reference or text books for researchers, academics and graduate students in Asian universities and research institutes, as well as in the West. With the growing strength of Asian economic, scientific and technological development, there is a need more than ever before for teaching and research materials written by leading Asian researchers, or those who have worked in or visited the Asian region, particularly tailored to meet the growing demands of student and researchers in that region. Many leading mathematicians in Asia were themselves trained in the West, and their experience with Western methods will make these books suitable not only for an Asian audience but also for the international mathematics community.

The *Asian Mathematics Series* is founded with the aim to present signficant contributions from mathematicians, written with an Asian audience in mind, to the mathematics community. The series will cover all mathematical fields and their applications, with volumes contributed to by international experts who have taught or performed research in Asia. The level of the material will be at graduate level or above. The book series will consist mainly of monographs and lecture notes but conference proceedings of meetings and workshops held in the Asian region will also be considered.

Preface

Periodic elastic plane problems are important in practice. There were many works in this field by various authors, most of which dealt with particular cases. However, it was lack of a systemetic treatment about this subject and even proper formulations for such problems did not occur in literature. The aim of this monograph is to make up this deficiency by means of methods of complex variables and illustrates the theory with various examples. Moreover, the general theory for analogous doubly periodic problems are also established but not in detail. Elements of complex analysis and linear elasticity are preassumed. The basic theory of boundary value problems for analytic functions and that of complex variable methods in elasticity (non-periodic case) may be found in, e.g, Lu [5] and [6] respectively. The most parts of the contents of this book are research works of the authors.

To make the book selfcontained, certain periodic boundary value problems for analytic functions are reminded in Chapter I, some parts of which may be found in Lu [5]. In Chapters II to VI, various periodic elastic problems are treated both in isotropic case and anisotropic case. The periodic crack problems are studied in Chapter V. A survey of doubly periodic elastic plane problems is illustrated in Chapter VI.

The book may be taken as a text for graduated students major in Applied Mathematics or Mechanics for one semester. It is also of benifit to scientific researchers and engineers in related fields.

In conclusion, it is a pleasure to us to acknowledge our indebtedness to World Scientific Publishing Co., Singapore, for its permission to use some materials from pp. 125 – 134 § 8, Chap. II of Lu [5].

Hai-tao Cai Jian-ke Lu

Chapter I

Periodic Boundary Value Problems for Analytic Functions

As a mathematical tool for investigation of periodic problems in plane elasticity, two kinds of fundamental boundary value problems for analytic functions will be discussed in the present chapter, namely, the periodic Riemann boundary value problems, the periodic Riemann-Hilbert boundary value problems in the half-plane. The more general formulations of these problems are those for automorphic functions which were studied by F. D. Gakhov and L. I. Chibrikova in the middle of this century. However, from the practical view-point, the periodic problems are much more important. The periodic problems discussed in this book are commonly confronted in applications.

More general results of such problems may be found in Lu [1] or [5].

§ 1. Periodic Riemann Boundary Value Problems: Case of Closed Contours

1. Formulation of the problems

Let L_k, $k=0, \pm 1, \pm 2, \cdots$, be a set of smooth closed contours, non-intersecting to each other, oriented counter-clockwisely, of the same shape and horizontally distributed with period $a\pi (a>0)$, as shown in Fig. 1.1.

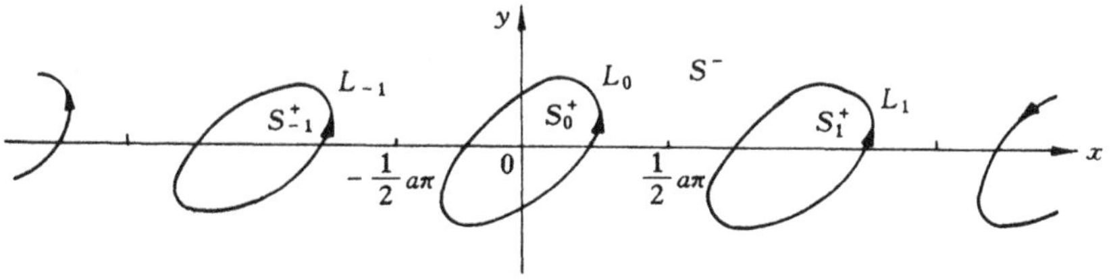

Fig. 1.1

The region interior to L_k is denoted by S_k^+ and that exterior to $L = \sum_{k=-\infty}^{+\infty} L_k$, by S^-. We may always assume the origin $O \in S_0^+$ and all the points $\pm\frac{1}{2}a\pi$, $\pm\frac{3}{2}a\pi$, $\cdots$ lie in S^-.

For briefness, we assume that L_k is a single closed contour. In fact, the following discussions remain effective with slight modifications when L_k consists of a finite number of non-interseting closed contours. Moreover, if L is an arbitrary smooth curve with period $a\pi$ (i.e., L_0 is a smooth arc from $-\frac{1}{2}a\pi + iy_0$ to $\frac{1}{2}a\pi + iy_0$ with the same slope at its end-points), they are also valid with suitable modifications.

Periodic Riemann boundary value problem (problem P_1): Find a function $\Phi(z)$ in the complex plane with period $a\pi$, satisfying

$$\Phi^+(t) = G(t)\Phi^-(t) + g(t), \quad t \in L, \tag{1.1}$$

where $\Phi^\pm(t)$ denote the boundary values (limiting values) of a *sectionally holomorphic function* $\Phi(z)$ (holomorphic in S_k, $k=0, \pm 1, \pm 2, \cdots$ and S^-) from the positive (left) and the negative (right) sides of L respectively, while $G(t)$ and $g(t) \in H$ are given on L with period $a\pi$:

$$G(t + a\pi) = G(t), \quad g(t + a\pi) = g(t), \quad t \in L,$$

and $G(t) \neq 0$ (*normal type*).

Here, a function $f(t) \in H$ (*Hölder continuous*) on L means

$$|f(t) - f(t')| \leqslant A|t - t'|^\mu, \quad \forall t, t' \in L,$$

for certain positive constants A and μ $(0 < \mu \leqslant 1)$.

Note that ∞ is a limiting point of the points on L so that the solution $\Phi(z)$ of the problem (if any) could not have a definite limit as $z \to \infty$ in general. However, we may ask $\Phi(z)$ meeting certain reguirements at $z = \pm\infty i$ (that means, for $z = x + iy$, x may be arbitrary and $y \to \pm\infty$ respectively). We always reguire $\Phi(\pm\infty i)$ to be bounded (i.e., finite).

The problem is *homogeneous* if $g(t) \equiv 0$, denoted by P_1^0, and otherwise, *non-homogeneous*.

Throughout this book, all periodic functions are always assumed to be of periodicity $a\pi (a > 0)$, for a definite a.

2. Transfer to classical Riemann boundary value problems

Denote the strip region $|\mathrm{Re}z| < \frac{1}{2}a\pi$ by S_0. Assume L_0 lies entirely in S_0 for the time being. Denote $S_0^- = S^- \cdot S_0$ and the positive directions of the straight lines $x = \pm\frac{1}{2}a\pi$ such that S_0^- situates in their right sides (Fig. 1.2). If

$\Phi(z)$ is a solution of the problem P_1, then, denoting its part in $S_0 = S_0^+ + S_0^-$ by $\Phi_0(z)$, we see that $\Phi_0(z)$ is a sectionally holomorphic function in S_0, continuous to $x = \pm\frac{1}{2}a\pi$ and satisfying

$$\Phi_0^+(t) = G(t)\Phi_0^-(t) + g(t),\ t \in L_0, \tag{1.2}$$

$$\Phi_0(\frac{1}{2}a\pi + iy) = \Phi_0(-\frac{1}{2}a\pi + iy),\ |y| < +\infty. \tag{1.3}$$

Conversely, if $\Phi_0(z)$ is a sectionally holomorphic function in S_0, continuous to $z = \pm\frac{a\pi}{2}$ and satisfying (1.2), (1.3), then, after its extension with period $a\pi$, a solution $\Phi(z)$ of the original problem P_1 is obtained.

In general, it is possible that L_0 may partially stepover the boundary of the strip $|\mathrm{Re}z| < \frac{1}{2}a\pi$ (Fig. 1.3). In such case, we may replace a pair of congruent line-segments (mod $a\pi$) on $x = \pm\frac{a\pi}{2}$ by two congruent smooth arcs such that S_0^+ would lie entirely in the interior of the modified curved strip S_0.

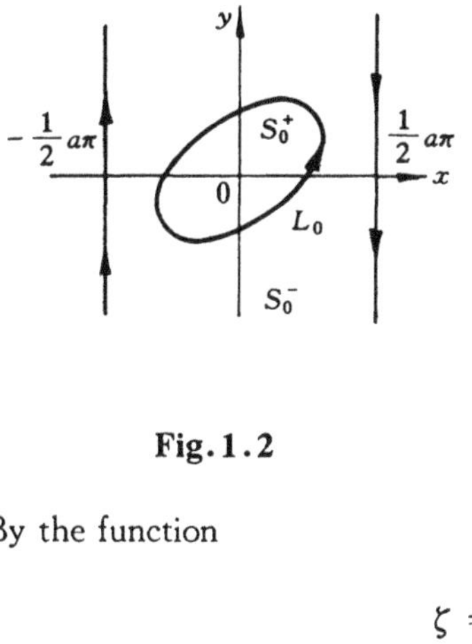

Fig. 1.2

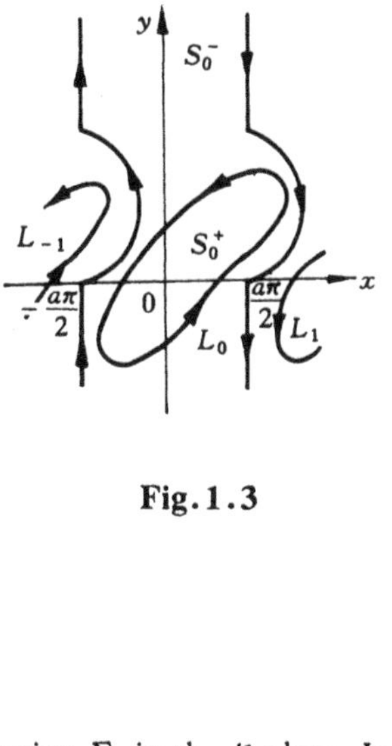

Fig. 1.3

By the function

$$\zeta = \tan\frac{z}{a}, \tag{1.4}$$

the strip region S_0 is conformally mapped to a region Σ_0 in the ζ-plane. In case of Fig. 1.2, it is the entire ζ-plane cut along the half-rays on the imaginary axis exterior to the interval $[-i, i]$, the points $z = 0, \pm\frac{1}{2}a\pi, +\infty i, -\infty i$ are mapped to $\zeta = 0, \infty, i, -i$ respectively, and the straight lines $x = \pm\frac{1}{2}a\pi$ become the left and the right banks of the cut respectively. In the mean time, L_0 is mapped into a certain smooth closed contour Γ_0, surrounding the origin O, with

$\zeta = \pm i$ in its exterior and non-intersecting with the cut (Fig. 1.4). In case of Fig. 1.3, the shape of the cut is changed, while Γ_0 remains not passing through the cut (Fig. 1.5). All the properties described above keep unchanged. The interior region of Γ_0 is denoted by Σ_0^+ and the exterior region, by Σ_0^-.

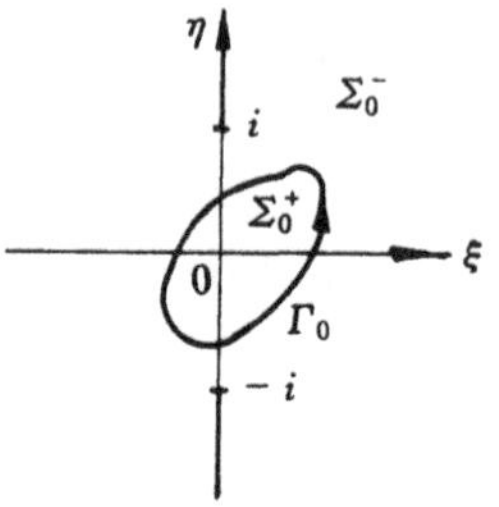

Fig. 1.4

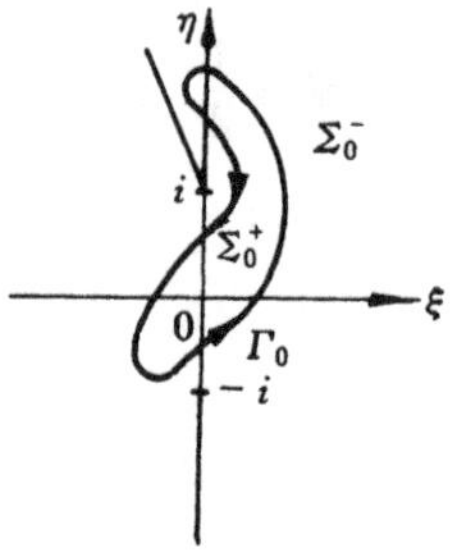

Fig. 1.5

Under the mapping, $\Phi_0(z)$ becomes a function $\Phi_*(\zeta)$, which has the same limiting values on the different sides of any point on the cut. Therefore, $\Phi_*(\zeta)$ is a sectionally holomorphic function in the ζ-plane (bounded at $\zeta = \infty$), satisfying

$$\Phi_*^+(\tau) = G_*(\tau)\Phi_*^-(\tau) + g_*(\tau), \quad \tau \in \Gamma_0, \tag{1.5}$$

where $G_*(\tau)$ and $g_*(\tau)$ are the image functions of $G(t)$ and $g(t)$ respectively under the mapping (1.4), both of them still $\in H$ and $G_*(\tau) \neq 0$. Thus, the problem is reduced to a classical Riemann problem. But we should pay attention to that $\zeta = \pm i$ may be isolated singular points of $\Phi_*(\zeta)$ in general.

The integer

$$k = \operatorname{Ind}_{L_0} G(t) = \frac{1}{2\pi}[\arg G(t)]_{L_0} \tag{1.6}$$

is called the *index* of the probem P_1, where $[f(t)]_{L_0}$ denotes the increment of $f(t)$ as t describes a complete cycle along the contour L_0 in its positive sense. It is actually the index of the transferred Riemann problem:

$$k = \operatorname{Ind}_{\Gamma_0} G_*(\tau)$$

since it is obvious $[\arg G(t)]_{L_0} = [\arg G_*(\tau)]_{\Gamma_0}$.

3. Discussion on homogeneous problem P_1^0

In this case, $g(t) \equiv 0$ which implies $g_*(\tau) \equiv 0$.

We have asked $\Phi(\pm\infty i)$ to be bounded. This means that $\Phi_*(\zeta)$ should be also bounded (and so, regular) at $\zeta = \pm i$. The general solution of (1.5) with $g_*(\tau) \equiv 0$ is

$$\Phi_*(\zeta) = X_*(\zeta)P_k(\zeta), \tag{1.7}$$

where

$$X_*(\zeta) = \begin{cases} e^{\Gamma_*(\zeta)}, & \zeta \in \Sigma_0^+, \\ \zeta^{-k}e^{\Gamma_*(\zeta)}, & \zeta \in \Sigma_0^-, \end{cases} \tag{1.8}$$

in which

$$\Gamma_*(\zeta) = \frac{1}{2\pi i}\int_{\Gamma_0} \frac{\log[\tau^{-k}G_*(\zeta)]}{\tau - \zeta}d\tau, \tag{1.9}$$

and $P_k(\zeta)$ is an arbitrary polynomial of degree (not greater than) k (regard $P_k(\zeta) \equiv 0$ when $k < 0$).

Returning to the z-plane, we have

$$\Phi_0(z) = X_0(z)P_k(\tan\frac{z}{a}), \tag{1.10}$$

where

$$X_0(z) = \begin{cases} e^{\Gamma(z)}, & z \in S_0^+, \\ \cot^k\frac{z}{a}e^{\Gamma(z)}, & z \in S_0^-, \end{cases} \tag{1.11}$$

in which

$$\begin{aligned} \Gamma(z) &= \frac{1}{2a\pi i}\int_{L_0} \frac{\log[\cot^k\frac{t}{a}G(t)]}{\tan\frac{t}{a} - \tan\frac{z}{a}} \frac{dt}{\cos^2\frac{t}{a}} \\ &= \frac{1}{2a\pi i}\int_{L_0} \log[\cot^k\frac{t}{a}G(t)](\cot\frac{t-z}{a} + \tan\frac{t}{a})dt \\ &= \frac{1}{2a\pi i}\int_{L_0} \log[\cot^k\frac{t}{a}G(t)]\cot\frac{t-z}{a}dt + C. \end{aligned}$$

Merging the constant C into P_k, we may regard

$$\Gamma(z) = \frac{1}{2a\pi i}\int_{L_0} \log[\cot^k\frac{t}{a}G(t)]\cot\frac{t-z}{a}dt. \tag{1.12}$$

The logarithm appeared above may be chosen as any continuous branch.

When $\Phi_0(z)$ is extended with period $a\pi$, since all the functions appeared on the right sides of (1.10) – (1.12) are periodic (with period $a\pi$), we get at once, without any change, the general solution of the original problem P_1^0:

$$\Phi(z) = X(z)P_k(\tan\frac{z}{a}), \tag{1.10$'$}$$

$$X(z) = \begin{cases} e^{\Gamma(z)}, & z \in S^+, \\ \cot^k \dfrac{z}{a} e^{\Gamma(z)}, & z \in S^-, \end{cases} \tag{1.11$'$}$$

where $\Gamma(z)$ is still given by (1.12). $X(z)$ is called the *characteristic function* of the problem P_1^0 (as well as P_1). The general solution (1.10)$'$ may be rewritten as

$$\Phi(z) = \begin{cases} \dfrac{1}{\cos^k \dfrac{z}{a}} e^{\Gamma(z)} Q_k\left(\sin \dfrac{z}{a}, \cos \dfrac{z}{a}\right), & z \in S^+, \\ \dfrac{1}{\sin^k \dfrac{z}{a}} e^{\Gamma(z)} Q_k\left(\sin \dfrac{z}{a}, \cos \dfrac{z}{a}\right), & z \in S^-, \end{cases} \tag{1.13}$$

where Q_k is an arbitrary homogeneous polynomial of degree k in its arguments ($Q_k \equiv 0$ when $k < 0$).

Thus, we obtain

Theorem 1.1 (Chibrikova) *If $\Phi^\pm(\infty i)$ are required to be bounded, then the homogeneous problem P_1^0 has $k+1$ linearly independent solutions when its index $k \geqslant 0$ and has only the trivial solution when $k < 0$.*

4. Discussion on non-homogeneous problem P_1

The following formula will be frequently used later:

Generalized Plemelj formula If $g(t) \in H$ with period $a\pi$ and

$$\Psi(z) = \frac{1}{2a\pi i} \int_{L_0} g(t) \cot \frac{t - z}{a} dt,$$

then,

$$\Psi^\pm(t_0) = \pm \frac{1}{2} g(t_0) + \frac{1}{2a\pi i} \int_{L_0} g(t) \cot \frac{t - t_0}{a} dt, \quad t_0 \in L, \tag{1.14}$$

where the integral appeared in the right-hand member is understood by the Cauchy principal value integral, called integral with *Hilbert kernel*.

For its proof, by the well-known Plemelj formula (Cf. Muskhelishvili [3] or Lu [5]), it is sufficient to seperate the principal part $\dfrac{a}{t - t_0}$ of $\cot \dfrac{t - t_0}{a}$ at $t = t_0$.

Let us now consider the non-homogeneous problem P_1(1.1), where $g(t) \not\equiv 0$ which gives rise to $g_*(\tau) \not\equiv 0$.

When $k \geqslant -1$, the general solution of (1.5) is

$$\Phi_*(\zeta) = X_*(\zeta)[\Psi_*(\zeta) + P_k(\zeta)],$$

where
$$\Psi_*(\zeta) = \frac{1}{2\pi i}\int_{\Gamma_0} \frac{g_*(\tau)}{X_*^+(\tau)}\frac{d\tau}{\tau - \zeta};$$

when $k < -1$, it is (uniquely) solvable iff the following $-k-1$ conditions are fulfilled:

$$\int_{\Gamma_0} \frac{g_*(\tau)}{X_*^+(\tau)}\tau^{j-1}d\tau = 0, \ j = 1, \cdots, -k-1.$$

Returning to the z-plane, we have, when $k \geqslant -1$, the general solution of P_1 is

$$\Phi(z) = X(z)[\Psi(z) + P_k(\tan\frac{z}{a})], \tag{1.15}$$

where

$$\Psi(z) = \frac{1}{2a\pi i}\int_{L_0} \frac{g(t)}{X^+(t)}(\cot\frac{t-z}{a} + \tan\frac{t}{a})dt. \tag{1.16}$$

When $k \geqslant 0$, the last term in the parentheses on its right side may be omitted by merging it into the constant term of P_k so that we may write

$$\Psi(z) = \frac{1}{2a\pi i}\int_{L_0} \frac{g(t)}{X^+(t)}\cot\frac{t-z}{a}dt. \tag{1.16$'$}$$

However, this term should not be omitted when $k = -1$ so as to guarantee $X(z)\Psi(z)$ to be bounded at $z = \pm\frac{1}{2}a\pi$. When $k < -1$, the problem P_1 has the unique solution (1.15) (with $P_k \equiv 0$) iff the following $-k-1$ conditions are fulfilled:

$$\int_{L_0} \frac{g(t)}{X^+(t)} \frac{\sin^{j-1}\frac{t}{a}}{\cos^{j+1}\frac{t}{a}}dt = 0, \ j = 1, \cdots, -k-1, \tag{1.17}$$

and $\Psi(z)$ should be given by (1.16) instead of (1.16)′.

Thus, we obtain

Theorem 1.2 (Chibrikova) *If $\Phi(\pm\infty i)$ are required to be bounded, then, the general solution of the non-homogeneous problem P_1 contains $k+1$ arbitrary constants when $k \geqslant -1$; it is (uniquely) solvable when $k < -1$ iff $-k-1$ conditions in (1.17) are fulfilled.*

5. A particular case

Consider the particular case $G(t) \equiv K$ being a constant ($K \neq 0$), which usually occurs in applications.

In this case, the index $k = 0$. Then,

$$\Gamma(z) = \frac{1}{2a\pi i}\int_{L_0} \log K \cdot \cot\frac{t-z}{a}dt = \frac{\log K}{2\pi i}[\log\sin\frac{t-z}{a}]_{L_0}$$
$$= \begin{cases} \log K, \text{ when } z \in S^+, \\ 0, \text{ when } z \in S^-, \end{cases}$$

where $\log K$ may be taken definitely and arbitrarily. Thus, $X^+(z) = K$, $X^-(z) = 1$. Therefore, as $\Phi(\pm\infty i)$ are bounded, the general solution of this problem P_1 is

$$\Phi(z) = \begin{cases} \dfrac{1}{2a\pi i}\displaystyle\int_{L_0} g(t)\cot\frac{t-z}{a}dt + C, \ z \in S^+, \\ \dfrac{1}{K}[\dfrac{1}{2a\pi i}\displaystyle\int_{L_0} g(t)\cot\frac{t-z}{a}dt + C], \ z \in S^-, \end{cases} \tag{1.18}$$

which may be checked directly by the generalized Plemelj formula (1.14).

§ 2. Periodic Riemann Boundary Value Problems: Case of Open Arcs or Discontinuous Coefficients

1. Case of open arcs

Consider the case $L = \sum_{k=-\infty}^{\infty} L_k$ being periodic as before where L_0 consists of p non-inter-secting smooth open arcs $a_r b_r$, $r = 1, \cdots, p$, oriented from a_r to b_r, and both $G(t)$, $g(t) \in H$ on each arc (including its end-points) with $G(t) \neq 0$.

Denote a_r, b_r, $r = 1, \cdots, p$, in uniformity as $c_1, \cdots, c_{2p}$. As in classical case (Cf. Lu[5]), they may be divided into two different kinds as follows. Take any continuous branch of $\log G(t)$ on each arc $a_r b_r$ and assume

$$\mp\frac{1}{2\pi i}\log G(c_j) = \alpha_j + i\beta_j, \ j = 1, \cdots, 2p,$$

where the negative (positive) sign is taken when $c_j = a_r (= b_r)$. If α_j is an integer, then c_j is called a *special end*, and otherwise, an *ordinary end*. Make arrangements such that $c_1, \cdots, c_m$ are ordinary ends and $c_{m+1}, \cdots, c_{2p}$ are special ones (either of them may be empty, i.e., $m = 2p$ or $m = 0$).

Let us find the periodic sectionally holomorphic function satisfying (1.1) (problem P_1), which keeps bounded in the neighborhoods of the ordinary ends $c_1, \cdots, c_q (q \leqslant m)$ and integrably unbounded (at most) near the remaining ends. Such solution of P_1 is said belonging to the *class* $h_q = h(c_1, \cdots, c_q)$. As in the classical case, it may be proved that, such a solution $\Phi(z)$ (if any) must be almost bounded near $c_{m+1} \cdots, c_{2p}$, i.e., $\lim_{z \to c_j}(z - c_j)^{\varepsilon}\Phi(z) = 0 (j > m)$ for any small $\varepsilon > 0$. If the solution is allowed to be integrably unbounded near all the end-points $c_1, \cdots, c_{2m}$, then it is said belonging to *class* h_0.

Define the index of problem (1.1) with respect to class h_q as follows. For any special end c_j, let $\lambda = -\alpha_j$; for an ordinary end c_j, choose integer λ_j such that $0 < \alpha_j + \lambda_j < 1$ when $j \leqslant q$ and $-1 < \alpha_j + \lambda_j < 0$ for $j > q$. Then,

$$k = -\sum_{j=1}^{2p} \lambda_j \tag{1.19}$$

is called the *index of* (1.1) *with respect to class* h_q.

By using the mapping (1.4) to the ζ-plane again, the problem is transferred to (1.5). Because L_0 does not pass through any singular point of (1.4), so $G_*(\tau)$ and $g_*(\tau)$ retain all the original properties on the image Γ_0 of L_0(Γ_0 consisting of a finite number of non-inter-secting smooth arcs too) and the type of each of its end-points remains unchanged. Thus, solution of the original problem (1.1) in class $h(c_1, \cdots, c_q)$ is transferred to that of (1.5) in class $h(c_1^*, \cdots, c_q^*)$, where we have denoted $c_j^* = \tan \frac{c_j}{a}$, $j = 1, \cdots, 2p$. As the solutions remain arbitrary but finite at $\zeta = \pm i$, $\Phi_*(\zeta)$ has to be bounded at infinity.

It is easily verified that, for homogeneous problem P_1^0, as the same as the properties of the solutions of problem (1.5) with $g_*(\tau) \equiv 0$, a solution $\Phi(z)$ is always bounded near its special ends and equals zero (if bounded) at its ordinary ones.

The general solution of (1.5) (with $g_*(\tau) \equiv 0$) in class h_q has the form

$$\Phi_*(\zeta) = X_*(\zeta) P_k(\zeta),$$

where

$$X_*(\zeta) = \Pi_*(\zeta) e^{\Gamma_*(\zeta)}$$

in which

$$\Pi_*(\zeta) = \prod_{j=1}^{2p} (\zeta - c_j^*)^{\lambda_j}, \quad \Gamma_*(\zeta) = \frac{1}{2\pi i} \int_{\Gamma_0} \frac{\log G_*(\tau)}{\tau - \zeta} d\tau.$$

Returning to the z-plane, we get the general solution of problem P_1^0 in class h_q:

$$\Phi(z) = X(z) P_k(\tan \frac{z}{a}), \tag{1.20}$$

where the characteristic function

$$X(z) = \Pi(z) e^{\Gamma(z)}, \tag{1.21}$$

in which

$$\Pi(z) = \prod_{j=1}^{2p} (\tan \frac{z}{a} - \tan \frac{c_j}{a})^{\lambda_j}, \tag{1.22}$$

$$\Gamma(z) = \frac{1}{2a\pi i} \int_{L_0} \log G(t) \tan \frac{t - z}{a} dt. \tag{1.23}$$

When a constant factor is merged into P_k, the general solution (1.20) may be rewritten as

$$\Phi(z) = \prod_{j=1}^{2p} \sin^{\lambda_j} \frac{z - c_j}{a} e^{\Gamma(z)} Q_k(\sin \frac{z}{a}, \cos \frac{z}{a}). \tag{1.20$'$}$$

Note that Theorem 1.1 in the previous section remains valiad in this case.[①]

We may analogously discuss solution of the so-called adjoint periodic boundary value problems in adjoint class, the results may be also easily deduced from the corresponding results in non-periodic classical case, which will be omitted here.

As in classical case, for non-homogeneous periodic problem P_1, its solution keeps bounded near the special end c_j if $G(c_j) \neq 1$ and is almost bounded if $G(c_j) = 1$ unless $g(c_j) = 0$.

2. An important special case

For the need in applications later, consider the special case where L_0 consists of a single line-segment $\gamma_0: -l \leqslant t \leqslant l$ $(l < \frac{1}{2} a\pi)$ along the real axis and $G(t) = -K$ is a negative real constant. We would seek for solutions in class h_0 of the corresponding problem P_1^0, i.e., solutions permitted to be integrably unbounded at most near $z = \pm l$.

Now $c_1 = -l$, $c_2 = l$. Take

$$\log(-K) = \ln K + \pi i$$

where $\ln K$ (always) denotes the real logarithm. Then

$$\alpha_1 + i\beta_1 = -\frac{1}{2} + i\beta, \quad \alpha_2 + i\beta_2 = \frac{1}{2} - i\beta,$$

where

$$\beta = \frac{\ln K}{2\pi}. \tag{1.24}$$

It is easily seen that

$$\Gamma(z) = \frac{\log(-K)}{2a\pi i} \int_{-l}^{l} \cot \frac{t - z}{a} dt = (\frac{1}{2} - i\beta) \log \frac{\tan \frac{z}{a} - \tan \frac{l}{a}}{\tan \frac{z}{a} + \tan \frac{l}{a}}$$

and so

① Of course, solutions are understood to be in class h_q and k, the index of the problem with respect to the same class.

$$X(z) = \Pi(z)e^{\Gamma(z)} = \left(\tan\frac{z}{a} + \tan\frac{l}{a}\right)^{-\frac{1}{2}+i\beta}\left(\tan\frac{z}{a} - \tan\frac{l}{a}\right)^{-\frac{1}{2}-i\beta}, \tag{1.25}$$

provided that the z-plane is cut along γ_0 and its congruent segments, and $X(z)$ is taken as a definite continuous branch arbitrarily, e.g.,

$$\lim_{z\to\pm\frac{1}{2}a\pi} \tan\frac{z}{a}X(z) = 1. \tag{1.26}$$

Thus, as $\Phi(\pm\infty i)$ are required to be bounded, the general solution of the problem P_1 is

$$\Phi(z) = \frac{X(z)}{2a\pi i}\int_{-l}^{l}\frac{g(t)}{X^+(t)}\cot\frac{t-z}{a}dt + X(z)\left(C_0\tan\frac{z}{a} + C_1\right). \tag{1.27}$$

In particular, if $G(t) = -1$, i.e., $K = 1$, then $\beta = 0$ by (1.24) and

$$X(z) = \frac{1}{i\sqrt{R(z)}}, \qquad R(z) = \tan^2\frac{l}{a} - \tan^2\frac{z}{a}, \tag{1.25$'$}$$

where $\sqrt{R(z)}$ has been taken as the branch such that it takes positive values as z tends to points on γ_0 from the upper half-plane.

In this subcase, the general solution is

$$\Phi(z) = \frac{1}{2a\pi i\sqrt{R(z)}}\int_{-l}^{l} g(t)\sqrt{R(t)}\cot\frac{t-z}{a}dt + \frac{C_0\tan\frac{z}{a} + C_1}{\sqrt{R(z)}}. \tag{1.27$'$}$$

3. Case of discontinuous coefficients

Assume L_0 to be a closed contour as in §1, but $G(t)$ and $g(t)$ have a finite number of discontinuities $c_1, \cdots, c_p$ of the first kind on L_0, belong to H on each "continuous" arc of L_0 (arc between two adjacent discontinuities) including its end-points (the values at these points are understood by the one-sided limits) and $G(t) \neq 0$.

As usual, call c_j an *ordinary node* when $G(c_j+0)/G(c_j-0)$ is not an integer, and otherwise, *a special node*. Denote the set of ordinary nodes by $\{c_1, \cdots, c_m\}$ and that of special nodes by $\{c_{m+1}, \cdots, c_p\}$. We want search solutions of problem P_1(1.1) in class $h_q = h(c_1, \cdots, c_q)$ $(q \leqslant m)$, i.e., solutuions bounded in the neighborhoods of $c_1, \cdots, c_q$ and (at most) integrably unbounded near the remaining nodes. Class h_0 is understood as before.

Define the index with respect to class h_q similarly as before, or, in our case: starting from a certain node, describing along the positive sense of L_0, we take an arbitrary continuous branch of $\log G(t)$ on the continuous arc following this node, and when passing through each c_j, we take a branch of $\log G(t)$ on the next arc with th value $\log G(c_j+0)$ such that

$$-\frac{1}{2\pi i}[\log G(c_j+0)-\log G(c_j-0)]=\alpha_j+i\beta_j, \tag{1.28}$$

where $\alpha_j=0$ if c_j is a special node, and, if c_j is an ordinary node, $0<\alpha_j<1$ when $j\leqslant q$ and $-1<\alpha_j<0$ when $j>q$. Repeat this process until returning to the positive side of the starting node. Then,

$$k=\frac{1}{2\pi i}[\log G(t)]_{L_0}=\frac{1}{2\pi}[\arg G(t)]_{L_0} \tag{1.29}$$

(including the jumps of $\log G(t)$ at all of the discontinuities) is called *the index of problem* (1.1) *with respect to class* h_q, which is obviously independent of the choice of the starting node.

After the notion of index is thus defined, all the discussions as well as the formulas given in § 1 are in effect in our case, which will not be repeated here.

Remark By the method used here, the so-called periodic Riemann boundary value problems with shift (Cf. Litvinchuk [1]) may be solved similarly.

§ 3. Periodic Riemann-Hilbert Boundary Value Problems of the Half-plane

1. Formulation of the problem

Assume L_k is the segment from $-\frac{1}{2}a\pi+\frac{1}{2}ka\pi$ to $\frac{1}{2}a\pi+\frac{1}{2}ka\pi$ ($k=0, \pm 1, \pm 2, \cdots$) lying on the real axis with length $a\pi$ ($a>0$). Then $\{L_k\}$ is a set of periodic segments with period $a\pi$, the union of which is the real axis. Denote the lower half-plane by S^- and the upper, by S^+.

The Riemann-Hilbert boundary value problem[①] *of the half-plane is : find a* function
$W(z)=u-iv$ holomorphic in S^- with period $a\pi$, satisfying the boundary condition

$$a(x)u+b(x)v=F(x),\ x\in L_k,\ k=0,\pm 1,\pm 2,\cdots, \tag{1.30}$$

where $a(x)$, $b(x)$ and $F(x)$ are given functions with period $a\pi$, arcwisely Hölder continuous, that is, having a finite number of discontinuities on each L_k, $\in H$ on each "continuous" closed interval between two adjacent discontinuous points (called nodes), and $a(x)$, $b(x)$ have no common zeros for any x. Without loss of generality, we may always assume that all of them are continuous at $x=\pm\frac{1}{2}a\pi$ (and their congruent points).

---

① Some authors call such a problem the Hilbert boundary value problem.

In some practical problems, certain restrictions are made on the behaviour of $W(z)$ on L_k near its singularities. For definiteness, we would solve (1.30) in the "wildest" class, that is, $W(z)$ may have integrable singularities at the nodes and it is bounded at $z=-\infty i$.

(1.30) is called *homogeneous* if $F(x)\equiv 0$ and otherwise, *non-homogeneous*. The general solution of (1.30) is of the form

$$W(z) = W_0(z) + W_1(z), \tag{1.31}$$

where $W_0(z)$ is the general solution of its corresponding homogeneous problem and $W_1(z)$ is its any particular solution, even bounded at the nodes.

2. Sketch of the method of solution

(1.30) may be written as

$$\mathrm{Re}\{[a(x)+ib(x)]W^-(x)\} = F(x),\ x \in L_0 \tag{1.32}$$

(the same for $x \in L_k$, $k=\pm 1, \pm 2, \cdots$, by periodicity), where $W^-(x)$ is the boundary value of $W(z)$ as $z \to x$ from S^-, or, what is the same,

$$[a(x)+ib(x)]W^-(x) + [a(x)-ib(x)]\overline{W^-(x)} = 2F(x),\ x \in L_0. \tag{1.32$'$}$$

Define a periodic function in S^+ by

$$\overline{W}(z) = \overline{W(\bar{z})},\ z \in S^+. \tag{1.33}$$

It is easily verified that $\overline{W}(z)$ is holomorphic in S^+ and its boundary value

$$\overline{W}^+(x) = \overline{W^-(x)},\ x \in L_0,$$

as $z \to x$ from S^+. Then we define a periodic and sectionally holomorphic function

$$\Omega(z) = \begin{cases} \overline{W}(z), & z \in S^+, \\ W(z), & z \in S^-, \end{cases} \tag{1.34}$$

with the x-axis as its jumping curve. For any function $f(z)$ ($\mathrm{Im}z \neq 0$), we may define $\bar{f}(z)=\overline{f(\bar{z})}$. It is easy to prove

$$\bar{\Omega}(z) = \Omega(z),\ \mathrm{Im}z \neq 0, \tag{1.35}$$

and consequently

$$\bar{\Omega}^{\pm}(x) = \overline{\Omega^{\mp}(x)},\ x \in L_0.$$

Thus, (1.32)′ may be rewritten as

$$\Omega^+(x) = -\frac{a(x)+ib(x)}{a(x)-ib(x)}\Omega^-(x) + \frac{2F(x)}{a(x)-ib(x)},\ x \in L_0, \tag{1.36}$$

which is a periodic Riemann boundary value problem of normal type as $a(x) \pm ib(x) \neq 0$ on L_0.

We may define, as before, the ordinary and the special nodes, the solution in class h_q and the index k with respect to this class. Now, our problem has been transferred to solve (1.36) in class h_0. Since $W(-\infty i) = \Omega(-\infty i)$ is required to be bounded and

$$\Omega(+\infty i) = \overline{\Omega(-\infty i)} \tag{1.37}$$

so that $\Omega(+\infty i)$ must be also bounded.

However, a solution of (1.36) is a solution of (1.32) (in the same class) iff (1.35) is fulfilled, and then the solution of the latter would be given by $\Omega^-(z)$, i.e., $\Omega(z)$ when $z \in S^-$. Note that (1.37) is actually true when (1.35) is valid. Thus, our problem (1.30) is equivalent to the problem (1.36) with supplementary condition (1.35) and $\Omega(\pm\infty i)$ to be bounded, which may be solved by the method as shown in the previous paragraph.

3. An important particular case

We would give solutions for a particular but very important case which often occur in practice and would be met later in this book.

Let $\gamma_0 = [-l, l]$, $0 < l < \frac{a\pi}{2}$, and $\gamma_0' = L_0 - \gamma_0$. Denote the set of all the segments congruent to γ_0 and γ_0' (mod $a\pi$) by γ and γ' respectively. In the particular case considered hereby:

$$a(x) + ib(x) = \begin{cases} a_1 + ib_1, & x \in \gamma, \\ a_2 + ib_2, & x \in \gamma', \end{cases} \tag{1.38}$$

where a_j, b_j are real constants with $a_j + ib_j \neq 0$ $(j = 1, 2)$ and $a_1 + ib_1 \neq a_2 + ib_2$.

Define single-valued function $\omega(x)$ with $-\pi < \omega(x) \leqslant \pi$ by

$$\cos\omega(x) = \frac{a(x)}{\sqrt{a(x)^2 + b(x)^2}}, \quad \sin\omega(x) = \frac{b(x)}{\sqrt{a(x)^2 + b(x)^2}}, \tag{1.39}$$

and write $\omega(x) = \omega_1$ when $x \in \gamma$ and $\omega(x) = \omega_2$ when $x \in \gamma'$. Then, $\omega_1 \neq \omega_2$ and

$$-\frac{a(x) + ib(x)}{a(x) - ib(x)} = e^{2i\omega(x) + i\pi}. \tag{$*$}$$

Let

$$\begin{aligned} \Gamma(z) &= \frac{1}{2a\pi i}\int_{L_0} [2i\omega(t) + i\pi]\cot\frac{t - z}{a}dt \\ &= \frac{1}{a\pi}\int_{L_0} [\omega(t) + \frac{\pi}{2}]\cot\frac{t - z}{a}dt, \quad \mathrm{Im}z \neq 0. \end{aligned} \tag{1.40}$$

Evidently

$$\bar{\Gamma}(z) = \Gamma(z) \tag{1.41}$$

and, by the generalized Plemelj formula,

$$\Gamma^+(x) - \Gamma^-(x) = 2i\omega(x) + i\pi \quad (x:\text{real}),$$

and so

$$e^{\Gamma^+(x)}/e^{\Gamma^-(x)} = -e^{2i\omega(x)} = -\frac{a(x) + ib(x)}{a(x) - ib(x)}. \tag{1.42}$$

We may evaluate $\Gamma(z)$ in finite form. We note that

$$\frac{1}{a\pi i}\int_{L_0} \cot\frac{t-z}{a}dt = \pm 1, \ z \in S^\pm, \tag{1.43}$$

which may be verified by the residue theorem (or the Cauchy theorem) for the infinite half-strip subregion of S^+ or S^+ with $|\mathrm{Re}z| < \frac{1}{2}a\pi$ and the fact $\cot\frac{t-z}{a} \to \pm i$ as $z \to \pm\infty i$. Then, (1.40) is actually

$$\Gamma(z) = \frac{1}{a\pi}\left\{\omega_1\int_{\gamma_0}\cot\frac{t-z}{a}dt + \omega_2\int_{\gamma_0'}\cot\frac{t-z}{a}dt\right\} \pm \frac{i\pi}{2}, \ z \in S^\pm. \tag{1.40$'$}$$

To evaluate the integral along γ_0 appeared in (1.40)′, let the z-plane be cut by γ. Then,

$$\frac{1}{a}\int_{\gamma_0}\cot\frac{t-z}{a}dt = \left[\log\sin\frac{t-z}{a}\right]_{\gamma_0} = \left\{\log\frac{\sin\frac{l-z}{a}}{\sin\frac{-l-z}{a}}\right\}_1$$

$$= \left\{\log\frac{\tan\frac{z}{a} - \tan\frac{l}{a}}{\tan\frac{z}{a} + \tan\frac{l}{a}}\right\}_1 = \ln\left|\frac{\tan\frac{l}{a} - \tan\frac{z}{a}}{\tan\frac{l}{a} + \tan\frac{z}{a}}\right| + i\theta_1(z), \tag{1.44}$$

where the branch of the logarithm involved with subscript 1 is taken such that

$$\theta_1(z) = \arg_1\left[\frac{\tan\frac{z}{a} - \tan\frac{l}{a}}{\tan\frac{z}{a} + \tan\frac{l}{a}}\right] \in \begin{cases}(0, \pi), \text{ when } z \in S^+, \\ (-\pi, 0), \text{ when } z \in S^-,\end{cases} \tag{1.45}$$

so that $\theta_1(\bar{z}) = -\theta_1(z)$,

$$\theta_1^\pm(x) = \begin{cases} \pm\pi, & \text{when } x \in \gamma, \\ 0, & \text{when } x \in \gamma'. \end{cases} \tag{1.46}$$

If we take the conformal mapping $\zeta = \tan\dfrac{z}{a}$, then $\theta_1(z)$ is the angle as shown in Fig. 1.6, where $L = \tan\dfrac{l}{a}$. ①

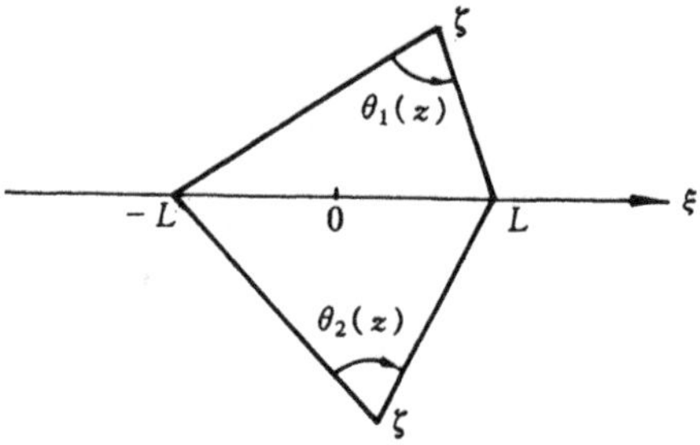

Fig. 1.6

Similarly, cutting the z-plane by γ', we have

$$\frac{1}{a}\int_{\gamma_0}\cot\frac{t-z}{a}dt = \frac{1}{a}\int_l^{a\pi-l}\cot\frac{t-z}{a}dt = [\log\sin\frac{t-z}{a}]_l^{a\pi-l}$$

$$= \left\{\log\frac{\sin\dfrac{a\pi-l-z}{a}}{\sin\dfrac{l-z}{a}}\right\}_2 = \left\{\frac{\tan\dfrac{l}{a}+\tan\dfrac{z}{a}}{\tan\dfrac{l}{a}-\tan\dfrac{z}{a}}\right\}_2$$

$$= \ln\left|\frac{\tan\dfrac{l}{a}+\tan\dfrac{z}{a}}{\tan\dfrac{l}{a}-\tan\dfrac{z}{a}}\right| + i\theta_2(z), \tag{1.47}$$

where

$$\theta_2(z) = \arg_2\left\{\frac{\tan\dfrac{l}{a}+\tan\dfrac{z}{a}}{\tan\dfrac{l}{a}-\tan\dfrac{z}{a}}\right\} \in \begin{cases} (0,\pi), & \text{when } z \in S^+, \\ (-\pi, 0), & \text{when } z \in S^-, \end{cases} \tag{1.48}$$

so that $\theta_2(\bar{z}) = -\theta_2(z)$,

$$\theta_2^\pm(x) = \begin{cases} 0, & x \in \gamma, \\ \pm\pi, & x \in \gamma'. \end{cases} \tag{1.49}$$

It is readily verified that

---

① When $l \to \frac{a\pi}{2}$, we have $L \to +\infty$ and $\theta_1(z) \to \pm\pi$ for $z \in S^\pm$ respectively, so that we get another proof of (1.43).

$$\theta_1(z) = \begin{cases} \pi - \theta_2(z), & \text{when } z \in S^+, \\ -\pi - \theta_2(z), & \text{when } z \in S^-. \end{cases} \tag{1.50}$$

Thus,

$$\left\{\log \frac{\tan \frac{z}{a} - \tan \frac{l}{a}}{\tan \frac{z}{a} + \tan \frac{l}{a}}\right\}_1 = \pm \pi i - \left\{\log \frac{\tan \frac{l}{a} + \tan \frac{z}{a}}{\tan \frac{l}{a} - \tan \frac{z}{a}}\right\}_2, \quad z \in S^\pm.$$

Substituting (1.43) and (1.46) into (1.40), and regarding the above equality, we obtain

$$\Gamma(z) = \pm \frac{\pi i}{2} \pm \omega_1 i + \frac{\omega_2 - \omega_1}{\pi}\left\{\log \frac{\tan \frac{l}{a} + \tan \frac{z}{a}}{\tan \frac{l}{a} - \tan \frac{z}{a}}\right\}_2$$

and hence

$$e^{\Gamma(z)} = \pm\, i e^{\pm \omega_1 i}\left\{\frac{\tan \frac{l}{a} + \tan \frac{z}{a}}{\tan \frac{l}{a} - \tan \frac{z}{a}}\right\}^\nu, \quad z \in S^\pm, \tag{1.51}$$

where we have put

$$\nu = \frac{\omega_2 - \omega_1}{\pi}, \quad 0 < |\nu| < 2, \tag{1.52}$$

and

$$\left\{\frac{\tan \frac{l}{a} + \tan \frac{z}{a}}{\tan \frac{l}{a} - \tan \frac{z}{a}}\right\}^\nu = \exp\left\{\nu \left\{\log \frac{\tan \frac{l}{a} + \tan \frac{z}{a}}{\tan \frac{l}{a} - \tan \frac{z}{a}}\right\}_2\right\}, \tag{1.53}$$

which is periodic and holomorphic in the z-plane cut by γ'.

It is easy to check directly that (1.41) and (1.42) are actually fulfilled by $e^{\Gamma(z)}$ with this expression. However, it could not serve as the canonical function of (1.36) since it has

$$\left|\frac{\tan \frac{l}{a} + \tan \frac{z}{a}}{\tan \frac{l}{a} - \tan \frac{z}{a}}\right|^\nu$$

as a factor which has zeros or "poles" at the tips of γ. To construct the canonical function $X(z)$, we may multiply $e^{\Gamma(z)}$ by a factor to annihilate these zeros and "poles".

In case $0<\nu<1$, we put

$$X(z) = e^{\Gamma(z)}/(\tan\frac{l}{a} + \tan\frac{z}{a})$$
$$= \pm ie^{\pm\omega_1 i}/[(\tan\frac{l}{a} - \tan\frac{z}{a})^{\nu}(\tan\frac{l}{a} + \tan\frac{z}{a})^{1-\nu}], \quad z \in S^{\pm}, \tag{1.54}$$

where the function in the denominator on the right side is holomorphic in the z-plane cut by γ' and both its factors take positive values when $z = x \in \gamma$. In this case, note that $X(z)$ has a zero-point of order 1 at $z = \pm\frac{a\pi}{2}$. Hence, the general solution of the homogeneous problem corresponding to (1.36) is

$$\Omega_0(z) = X(z)(C_1\tan\frac{z}{a} + C_2), \tag{1.55}$$

where C_1 and C_2 are arbitrary constants.

Note that $X(z)$, as $e^{\Gamma(z)}$, fulfills $\overline{X}(z) = X(z)$. Therefore, in order to guarantee $\Omega(z)$ satisfying (1.35), C_1 and C_2 have to be taken as real.

In case $1<\nu<2$, we put

$$X(z) = e^{\Gamma(z)}\frac{\tan\frac{l}{a} - \tan\frac{z}{a}}{(\tan\frac{l}{a} + \tan\frac{z}{a})^2}$$
$$= \pm ie^{\pm\omega_1 i}/[(\tan\frac{l}{a} + \tan\frac{z}{a})^{2-\nu}(\tan\frac{l}{a} - \tan\frac{z}{a})^{\nu-1}]. \tag{1.56}$$

By the same reasoning, we also have (1.54) as the general solution of the homogeneous problem with real C_1 and C_2, provided that $X(z)$ is given by (1.56).

For the case $\nu = 1$, it is obvious that $-\frac{a(x)+ib(x)}{a(x)-ib(x)}$ is actually equal to a single constant all over the x-axis and the corresponding problem (1.36) has a constant coefficient, which is a simple case.

When $-2<\nu<0$, anologous discussions may be made.

A particular solution of (1.36) is, as well-known,

$$\Omega_1(z) = \frac{X(z)}{a\pi i}\int_{L_0}\frac{F(t)}{X^+(t)[a(t)-ib(t)]}\cot\frac{t-z}{a}dt, \quad \mathrm{Im}z \neq 0. \tag{1.57}$$

We would prove that (1.35) is fulfilled by $\Omega_1(z)$. In fact, we have

$$\overline{\Omega}_1(z) = -\frac{\overline{X}(z)}{a\pi i}\int_{L_0}\frac{\overline{F(t)}}{\overline{X^+(t)[a(t)+ib(t)]}}\cot\frac{t-z}{a}dt, \quad \mathrm{Im}z \neq 0.$$

It is already known that $\overline{X}(z) = X(z)$ and $X^+(t)/X^-(t) = -e^{2i\omega(t)} = -$

$\frac{a(t)+ib(t)}{a(t)-ib(t)}$ and so

$$\overline{X^+(t)} = -\frac{a(t)-ib(t)}{a(t)+ib(t)}\overline{X^-(t)} = -\frac{a(t)-ib(t)}{a(t)+ib(t)}\overline{X}^+(t)$$

or

$$\overline{X^+(t)}[a(t)+ib(t)] = -X^+(t)[a(t)-ib(t)].$$

Substituting it into the above equality, we readily obtain $\overline{\Omega}_1(z) = \Omega_1(z)$.

Thus, the solution of our problem (1.30) is, when writing $W(z) = u - iv$,

$$W(z) = \Omega(z) = \Omega_1(z) + \Omega_0(z),\ z \in S^-, \tag{1.58}$$

where $\Omega_1(z)$ and $\Omega_0(z)$ are given by (1.57) and (1.55) respectively.

§ 4. Some Integral Formulas for Hilbert Kernel in the Half-Plane

For later needs, we establish the following theorem.

Theorem 1.3 *For a function $g(z)$ holomorphic in the lower half-plane S^-, with period $a\pi$ $(a>0)$, continuous to the real axis and bounded at $z=-\infty i$, we have the following formulas:*

$$\frac{1}{2a\pi i}\int_{L_0} g(t)\cot\frac{t-z}{a}dt = -g(z) + \frac{1}{2}g(-\infty i),\ z \in S^-; \tag{1.59}$$

$$\frac{1}{2a\pi i}\int_{L_0} \overline{g(t)}\cot\frac{t-z}{a}dt = \frac{1}{2}\overline{g(-\infty i)},\ z \in S^-. \tag{1.60}$$

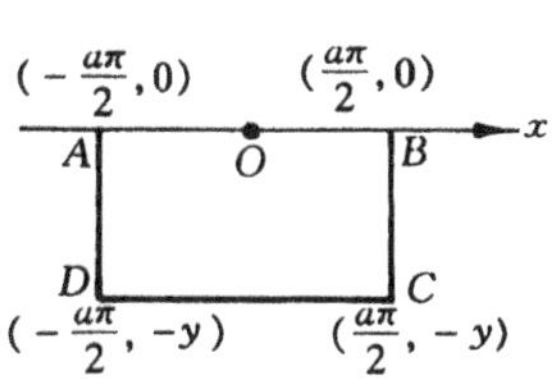

Fig. 1.7

Proof Integrating $g(t)\cot\frac{t-z}{a}$ along the boundary Γ of the rectangle $ABCD$ counterclockwisely, where $A=(-\frac{1}{2}a\pi,\ 0)$, $B=(\frac{1}{2}a\pi,\ 0)$, $C=(\frac{1}{2}$

$a\pi, -h)$, $D=(-\frac{1}{2}a\pi, -h)$ with $h>0$ sufficiently large such that z lies in the rectangle (Fig. 1.7), we obtain, by the residue theorem,

$$\frac{1}{2a\pi i}\int_{\Gamma} g(t)\cot\frac{t-z}{a}dt = -g(z). \tag{1.61}$$

The integral on the left side of (1.60) equals

$$\left\{\int_{AB} + \int_{BC} + \int_{CD} + \int_{DA}\right\} g(t)\cot\frac{t-z}{a}dt.$$

But

$$\left\{\int_{BC} + \int_{DA}\right\} g(t)\cot\frac{t-z}{a}dt = 0$$

by periodicity of $g(t)\cot\frac{t-z}{a}$, while the integral along CD tends to $a\pi i g(-\infty i)$ as $h\to+\infty$. Thus, (1.59) is readily proved.

(1.60) may be proved in a similar manner.

We may have analogous equations for $g(z)$ holomorphic in S^+.

This theorem is in fact an extension of (1.43).

Chapter Ⅱ

Periodic Problems for Isotropic Material in Plane Elastic Theory

There were many works in literature on the periodic problems in the theory of plane elasticity for isotropic material, e.g., researches on stress distribution in the neighborhoods of periodic holes by R. C. J. Howland, G. N. Savin, L. M. Tang, M. Isida, study on periodic contact problems by G. M. L. Gladwall and investigation for expression of periodic stress functions by S. Morigashi. All these works have limitations in either the shape of the holes or the boundary condition, and the discussions are incomplete. Especially, there were few studies on the possibility of *quasi-periodic displacements*.

In the present chapter, first of all, we would discuss the general expression of the stress functions of isotropic infinite elastic plane with periodically distributed holes of arbitrary shape. We would establish that, if the stresses are periodic, then the displacements are *quasi-periodic* in general, that is, they would get constant increments after a period, by which we may formulate the periodic first fundamental problem. We would also prove that the converse is valid too and then give the general formulation of the periodic second fundamental problem. However, for briefness of discussion, from § 2 on, we would assume the displacements are periodic, which usually meets the requirements in practical engineering, while discussions for the case of quasi-periodic displacements could not occur any difficulty in principle. In the remaining parts of the chapter, periodic welding problems, fundamental problems and contact problems will be considered.

§ 1. Stress Functions

1. General expression of stress functions

Let the elastic plane possess a row of periodic holes with boundaries L_j, $j = 0, \pm 1, \pm 2, \cdots$, where L_j consists of n arc-wisely smooth closed contours $l_k^{(j)}$, $k = 1, \cdots, n$ (Fig.2.1) and for the same k, $l_k^{(j)}$, $j = 0, \pm 1, \pm 2, \cdots$, are periodically arranged. $l_k^{(0)}$ will be denoted briefly by l_k. All the contours are oriented counter-clockwisely. The region occupied by the elastic body is denoted as S^-, the region bounded by $l_k^{(j)}$ as $S_k^{(j)+}$, and $S_k^{(0)+}$ briefly as S_k^+. Denote the strip region $|x|<\frac{1}{2}a\pi$ by S_0. Let $S_0^+=\sum_{k=1}^{n}S_k^+$ and $S_0^-=S_0-S_0^+$.

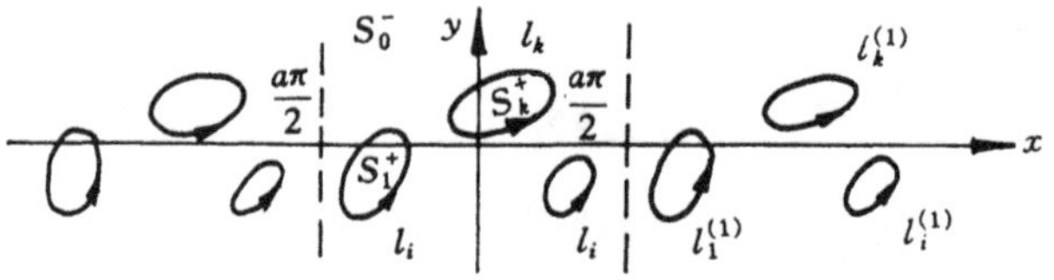

Fig. 2.1

Denote the stresses at any point $z=x+iy$ in S^- by $\sigma_x(z), \sigma_y(z), \tau_{xy}(z)$ and the (complex) displacement by $D(z)=u+iv$. It is well-known that they may be expressed in terms of (complex) *stress functions* $\varphi(z)$ and $\psi(z)$ or their derivatives $\Phi(z)=\varphi'(z)$ and $\Psi(z)=\psi'(z)$ as follows (Cf. Muskhelishvili [1]):

$$\sigma_x+\sigma_y=2[\Phi(z)+\overline{\Phi(z)}], \tag{2.1}$$

$$\sigma_y-\sigma_x+2i\tau_{xy}=2[\bar{z}\Phi'(z)+\Psi(z)], \tag{2.2}$$

$$2\mu D=2\mu(u+iv)=\kappa\varphi(z)-z\overline{\varphi'(z)}-\overline{\psi(z)}, \tag{2.3}$$

where μ is the shearing modulus of the elastic medium, κ is a constant related to its Poisson ratio σ ($1<\kappa<3$), and $\varphi(z)$, $\psi(z)$ are functions analytic (in general, multi-valued) functions in S^- with their derivatives holomorphic, i.e., single-valued, in S^-.

We always assume that the stresses are periodic and bounded at $z=\pm\infty i$.

Under this assumption, for the stress distribution, it is sufficient to know the external stresses along the boundary contours. Denote the principal vector of the external stresses on l_k by X_k+iY_k. Then the resultant on L_0 is

$$X+iY=\sum_{k=1}^{n}(X_k+iY_k). \tag{2.4}$$

Denote the stresses at $z=\pm\infty i$ respectively by

$$\sigma_\pm=\sigma_y(\pm\infty i),\quad \tau_\pm=\tau_{xy}(\pm\infty i),\quad h_\pm=\sigma_x(\pm\infty i). \tag{2.5}$$

Considering the equilibrium on the strip S_0^-, by periodicity of the stresses, we readily obtain

$$\sigma_--\sigma_+=\frac{Y}{a\pi},\qquad \tau_--\tau_+=\frac{X}{a\pi}. \tag{2.6}$$

Moreover, we note that, by (2.1), $\mathrm{Re}\Phi(z)$ is bounded at $z=\pm\infty i$. Let Φ

(z) become $\Phi_*(\zeta)$ by the mapping

$$\zeta = \tan\frac{z}{a}.$$

Then $\mathrm{Re}\Phi_*(\zeta)$ is bounded at $\zeta = \pm i$ and so they are regular points of $\Phi_*(\zeta)$, i.e., $\Phi(\pm\infty i)$ are finite.

Since

$$\Phi'(z) = \Phi'_*(\zeta)\frac{d\zeta}{dz} = \frac{\Phi'_*(\zeta)}{a\cos^2\frac{z}{a}},$$

we have

$$\lim_{z\to\pm\infty i} z\Phi'(z) = 0. \tag{2.7}$$

We also know that $\Psi(\pm\infty i)$ are finite by (2.2).

We want to get the general expression of $\varphi(z)$ and $\psi(z)$ such that their multi-valued parts and non-periodic parts are seperated out.

Take an arbitrary point z_k in S_k^+ for each k $(1\leqslant k\leqslant n)$. It is known that the multi-valued part of $\varphi(z)$ would get an increment

$$-\frac{X_k + iY_k}{2\pi(\kappa+1)}\log(z - z_k - ja\pi)$$

when z describes a cycle counter-clockwisely around the hole $S_k^{(j)+}$. In order to seperate out the whole multi-valued part of $\varphi(z)$, it is impossible to take summation of this expression directly for k and j since it would give rise to a divergent series. Because it would not influence the multi-valued property when the logarithmic factor in it is changed to

$$\log\left(1 - \frac{z - z_k}{ja\pi}\right) \quad (j \neq 0),$$

so, by using the infinite product representation of $\sin\frac{z - z_k}{a}$, it is readily seen that

$$\varphi(z) = -\frac{1}{2\pi(\kappa+1)}\sum_{k=1}^{n}(X_k + iY_k)\log\sin\frac{z - z_k}{a} + \varphi^*(z), \tag{2.8}$$

where $\varphi^*(z)$ is already holomorphic in S^-.

Similarly, we have

$$\psi(z) = \frac{\kappa}{2\pi(\kappa+1)}\sum_{k=1}^{n}(X_k - iY_k)\log\sin\frac{z - z_k}{a} + \psi^*(z), \tag{2.9}$$

where $\psi^*(z)$ is also holomorphic in S^-. Thereby,

$$\Phi(z) = -\frac{1}{2a\pi(\kappa+1)}\sum_{k=1}^{n}(X_k + iY_k)\cot\frac{z-z_k}{a} + \Phi^*(z), \quad (2.10)$$

$$\Psi(z) = \frac{\kappa}{2a\pi(\kappa+1)}\sum_{k=1}^{n}(X_k - iY_k)\cot\frac{z-z_k}{a} + \Psi^*(z), \quad (2.11)$$

in which we have put $\Phi^*(z) = \varphi^{*\prime}(z)$, $\Psi^*(z) = \psi^{*\prime}(z)$.

Applying Mikhlin's result ([1], § 55), by periodicity of stresses, we get the following relations:

$$\Phi(z + a\pi) = \Phi(z) + i\alpha,$$
$$\Psi(z + a\pi) = \Psi(z) - a\pi\Phi'(z),$$

where α is a certain real constant, which is actually zero since $\Phi(\pm\infty i)$ are finite. Therefore, we have correspondingly

$$\begin{aligned}\Phi^*(z + a\pi) &= \Phi^*(z),\\ \Psi^*(z + a\pi) &= \Psi^*(z) - a\pi\Phi'(z)\\ &= \Psi^*(z) - \frac{1}{2a(\kappa+1)}\sum_{k=1}^{n}\frac{X_k + iY_k}{\sin^2\frac{z-z_k}{a}} - a\pi\Phi^{*\prime}(z).\end{aligned}$$

Noting that both φ^* and ψ^* are single-valued, we obtain, by integration,

$$\left.\begin{aligned}\varphi^*(z + a\pi) &= \varphi^*(z) + a\pi\beta,\\ \psi^*(z + a\pi) &= \psi^*(z) - a\pi\varphi'(z) + \gamma\\ &= \psi^*(z) + \frac{1}{2(\kappa+1)}\sum_{k=1}^{n}(X_k + iY_k)\cot\frac{z-z_k}{a} - a\pi\Phi^*(z) + \gamma,\end{aligned}\right\} \quad (2.12)$$

where β and γ are certain complex constants.

Replacing z by $z + a\pi$ in (2.3), subtrating it from (2.3), substituting (2.9), (2.10) into the result and using (2.12), we have

$$2\mu[D(z + a\pi) - D(z)] = a\pi\kappa\beta - \bar{\gamma}.$$

By disregarding a rigid rotation of the elastic body, its right-hand member may be considered as a real constant.

Thus, we obtain the following theorem.

Theorem 2.1 *In the isotropic infinite elastic plane weakened by a row of holes of any shape with period $a\pi$ ($a > 0$), if the stresses are known to be periodic and bounded at infinity, then the relative displacements must be quasi-periodic, i.e.,*

$$\left.\begin{aligned}2\mu[u(z + a\pi) - u(z)] &= a\pi q,\\ v(z + a\pi) - v(z) &= 0,\end{aligned}\right| \quad (2.13)$$

where q is a certain real constant.

$\frac{a\pi q}{2\mu}$ is called the *addendum* of the *quasi-periodic function* $u(z)$.

By the previous expression, we know that

$$\gamma = a\pi(\kappa\bar{\beta} - q). \tag{2.14}$$

Thus, (2.12) becomes

$$\begin{aligned} \varphi^*(z + a\pi) &= \varphi^*(z) + a\pi\beta, \\ \psi^*(z + a\pi) &= \psi^*(z) - a\pi\varphi'(z) + a\pi(\kappa\bar{\beta} - q). \end{aligned}$$

Set

$$\left.\begin{aligned} \varphi_0(z) &= \varphi^*(z) - \beta z, \\ \psi_0(z) &= \psi^*(z) + z\varphi'(z) - (\kappa\bar{\beta} - q)z. \end{aligned}\right\} \tag{2.15}$$

It is easily seen that both $\varphi_0(z)$ and $\psi_0(z)$ are periodic functions holomorphic in S^-. β and q may be determined by the stresses at $z = \pm\infty i$, which will be seen later.

By (2.10) and (2.11), we have

$$\Phi(z) = -\frac{1}{2a\pi(\kappa+1)}\sum_{k=1}^{n}(X_k + iY_k)\cot\frac{z - z_k}{a} + \varphi_0'(z) + \beta, \tag{2.16}$$

$$\begin{aligned} \Psi(z) = {} & \frac{\kappa}{2a\pi(\kappa+1)}\sum_{k=1}^{n}(X_k - iY_k)\cot\frac{z - z_k}{a} + \psi_0'(z) \\ & + \frac{1}{2a\pi(\kappa+1)}\sum_{k=1}^{n}\left[(X_k + iY_k)\left(\cot\frac{z - z_k}{a} - \frac{z}{a\sin^2\frac{z - z_k}{a}}\right)\right] \\ & - \varphi_0'(z) - z\varphi_0''(z) - (\beta - \kappa\bar{\beta} + q). \end{aligned} \tag{2.17}$$

Both $\varphi_0'(z) = \Phi_0(z)$ and $\psi_0'(z) = \Psi_0(z)$ are bounded at infinity as $\Phi(\pm\infty i)$ and $\Psi(\pm\infty i)$ are finite, and then, by (2.8),

$$\lim_{z\to\pm\infty i} z\Phi_0'(z) = 0. \tag{2.18}$$

Furthermore, noting that $\varphi_0(z)$ is periodic, we know

$$0 = \frac{1}{a\pi}[\varphi_0(z + a\pi) - \varphi_0(z)] = \frac{1}{a\pi}\int_z^{z+a\pi}\Phi_0(z)dz = \Phi_0(\pm\infty i),$$

where the last equality is obtained as follows. Let the path of integration be the line-segment from z to $z + a\pi$ (it would lie entirely in S^- when $|z|$ is sufficiently large), seperate its real and imaginary parts, use the mean-value theorem respectively and let $z\to\pm\infty i$. Similarly, $\Psi_0(\pm\infty i) = 0$. And so both $\varphi_0(\pm\infty i)$ and $\psi_0(\pm\infty i)$ are finite.

Eliminating σ_x between (2.1) and (2.2), and letting $z\to\pm\infty i$, we have, by the above explanation and (2.16), (2.17),

$$\sigma_\pm + i\tau_\pm = \mp \frac{Y - iX}{2a\pi} + (\kappa + 1)\bar{\beta} - q.$$

By subtraction, we obtain equilibrium condition (2.6) again. By adding these two
equalities and taking conjugates, we get

$$(\kappa + 1)\beta - q = \frac{\sigma_- + \sigma_+}{2} - i\frac{\tau_- + \tau_+}{2}. \tag{2.19}$$

On the other hand, we have, by (2.16) and (2.17),

$$\Phi(\pm \infty i) = \mp \frac{Y - iX}{2a\pi(\kappa + 1)} + \beta.$$

Therefore, by (2.1), it gives rise at once

$$h_\pm + \sigma_\pm = \mp \frac{2Y}{a\pi(\kappa + 1)} + 2(\beta + \bar{\beta}), \tag{2.20}$$

by which it is seen that $h_\pm$ are not arbitrary and have to satisfy the condition

$$h_+ + \sigma_+ + \frac{2Y}{a\pi(\kappa + 1)} = h_- + \sigma_- - \frac{2Y}{a\pi(\kappa + 1)},$$

that is, by (2.6),

$$h_- - h_+ = \frac{3 - \kappa}{\kappa + 1}\frac{Y}{a\pi}. \tag{2.21}$$

At length, by (2.19) and (2.20), it is easy to evaluate

$$q = \frac{1}{4}[(\kappa + 1)h_+ - (3 - \kappa)\sigma_+] = \frac{1}{4}[(\kappa + 1)h_- - (3 - \kappa)\sigma_-]. \tag{2.22}$$

Substituting it into (2.19), we find

$$\begin{aligned}\beta &= \frac{1}{4}(h_+ + \sigma_+) + \frac{Y - iX}{2a\pi(\kappa + 1)} - \frac{i\tau_+}{\kappa + 1} \\ &= \frac{1}{4}(h_- + \sigma_-) - \frac{Y - iX}{2a\pi(\kappa + 1)} - \frac{i\tau_-}{\kappa + 1}.\end{aligned} \tag{2.23}$$

Thus, β and q are uniquely determined by the stresses at $z = \pm\infty i$ and the resultant of the principal vectors of the external stresses themselves along the boundary contours of the holes in a periodic strip, but independent of the external stresses on these boundary contours.

Substituting the above results in (2.8) and (2.9), we finally get the following theorem.

Theorem 2.2 *Suppose there is a row of holes with period* $a\pi$ $(a > 0)$ *in* *the isotropic elastic plane, the boundary of which is* L_j, $j = 0, \pm 1, \pm 2, \cdots$, *where each* L_j *consists of* n *arc-wisely smooth closed contours* $l_k^{(j)}$, $k = 1, 2, \cdots, n$, *and for a fixed* k, $l_k^{(j)}$, $j = 0, \pm 1, \pm 2, \cdots$, *are arranged periodically. Assume the stresses are* $a\pi$*-periodic, bounded at* $z = \pm\infty i$ *and the principal vector of the external stresses on* $l_k^{(0)}$ *is* $X_k + iY_k$. *Then, the stress functions* $\varphi(z)$ *and* $\psi(z)$ *respectively have the expressions*

$$\varphi(z) = -\frac{1}{2\pi(\kappa+1)}\sum_{k=1}^{n}(X_k + iY_k)\log\sin\frac{z - z_k}{a} + \beta z + \varphi_0(z), \tag{2.24}$$

$$\begin{aligned}\psi(z) = {} & \frac{\kappa}{2\pi(\kappa+1)}\sum_{k=1}^{n}(X_k - iY_k)\log\sin\frac{z-z_k}{a} \\ & + \frac{z}{2a\pi(\kappa+1)}\sum_{k=1}^{n}(X_k + iY_k)\cot\frac{z-z_k}{a} \\ & + (\kappa\bar{\beta} - \beta + q)z - z\varphi_0'(z) + \psi_0(z),\end{aligned} \tag{2.25}$$

where $\varphi_0(z)$ *and* $\psi_0(z)$ *are functions holomorphic and* $a\pi$*-periodic in the elastic region* S^- *and* q, β *are given by* (2.22), (2.23) *respectively.*

Thus, we have seperated out the multi-valued and non-periodic parts of φ, ψ. S. Morigashi had given their general expressions under the condition of periodic stresses but had not seperated out their multi-valued parts and had not assumed the boundedness of the stresses at infinity so that the connections among the constants occuring in these expressions disappeared.

Corollary 2.1 *Assume the elastic body as above. If the resultant of the principal vectors on the boundaries of the holes in a periodic strip is zero and the stresses at* $z = \pm\infty i$ *are* $\sigma = \sigma_y(\pm\infty i)$ *and* $\tau = \tau_{xy}(\pm\infty i)$,① *then both the stress functions* $\varphi(z)$ *and* $\psi(z)$ *are single-valued, and*

$$\varphi(z) = \varphi_0(z) + \beta z, \tag{2.26}$$

$$\psi(z) = \psi_0(z) - z\varphi'(z) + \kappa\bar{\beta}z = (\kappa\bar{\beta} - \beta)z - z\varphi_0'(z) + \psi_0(z), \tag{2.27}$$

where

$$\beta = \frac{-\sigma + i\tau}{1+\kappa}, \tag{2.28}$$

while both $\varphi_0(z)$ *and* $\psi_0(z)$ *are* $a\pi$*-periodic.*

2. The converse of Theorem 2.1

Theorem 2.3 *If the displacements in an isotropic plane elastic body are*

① In this case, the stresses at $z = \pm\infty i$ must be equal to each other so as the equilibrium is guaranteed.

quasi-periodic, then the stresses must be periodic.

Proof Assume the horizontal displacements are quasi-periodic and the vertical ones are periodic:

$$2\mu[u(z+a\pi)-u(z)]=a\pi q, \quad v(z+a\pi)=v(z). \tag{2.29}$$

Since

$$2\mu(u+iv)=\kappa\varphi(z)-z\overline{\varphi'(z)}-\overline{\psi(z)}, \tag{2.30}$$

we have,

$$\kappa[\varphi(z+a\pi)-\varphi(z)]-z[\overline{\varphi'(z+a\pi)}-\overline{\varphi'(z)}]-a\pi\overline{\varphi'(z+a\pi)} - [\overline{\psi(z+a\pi)}-\overline{\psi(z)}]=a\pi q$$

and, by differentiation,

$$\{\kappa[\Phi(z+a\pi)-\Phi(z)]-[\overline{\Phi(z+a\pi)}-\overline{\Phi(z)}]\}dz - \{z[\overline{\Phi'(z+a\pi)}-\overline{\Phi'(z)}]+a\pi\overline{\Phi'(z+a\pi)} + [\overline{\Psi(z+a\pi)}-\overline{\Psi(z)}]\}d\bar{z}=0.$$

In the expression $d\bar{z}=e^{i\theta}dz$, θ may be arbitrary as does dz, so that

$$\kappa[\Phi(z+a\pi)-\Phi(z)]=\overline{\Phi(z+a\pi)}-\overline{\Phi(z)}, \tag{2.31}$$

$$z[\Phi'(z+a\pi)-\Phi'(z)]+a\pi\Phi'(z+a\pi)+\Psi(z+a\pi)-\Psi(z)=0. \tag{2.32}$$

Taking conjugates in (2.31), substituting back and noting $\kappa>1$, we conclude

$$\Phi(z+a\pi)=\Phi(z). \tag{2.33}$$

Substituting it into (2.32), we get

$$\Psi(z+a\pi)-\Psi(z)=-a\pi\Phi'(z). \tag{2.34}$$

By combining (2.33) and (2.34), the periodicity of the stresses is established at once.

This theorem gives a solid foundation for solving second fundamental problems under the assumption of quasi-periodicity of the displacements.

Remark By the *generalized Hooke's law* (see Lu[6]),

$$\sigma_x=(\lambda+2\mu)\frac{\partial u}{\partial x}+\lambda\frac{\partial v}{\partial y},$$

$$\sigma_y=\lambda\frac{\partial u}{\partial x}+(\lambda+2\mu)\frac{\partial v}{\partial y},$$

$$\tau_{xy} = \mu\left(\frac{\partial v}{\partial x} + \frac{\partial u}{\partial y}\right),$$

where λ, μ are the Lamé constants of the elastic medium, it is readily seen that the stresses are periodic when the displacements are quasi-periodic. However, here we get something more such as (2.33) and (2.34).

3. Formulation of the fundamental problem

First fundamental problem Given the periodic stress function $X_n(t) + iY_n(t)$ on the boundary L of the elastic region and the stresses at $z = -\infty i$ (or $z = +\infty i$), find the elastic equilibrium (i. e., the stress distribution in the elastic body).

This is the most general formulation of the problem under the assumption of stresses to be periodic and bounded at infinity. In this case, by (2.6) and (2.21), the stresses at $z = +\infty i$ (or $z = -\infty i$) are also known, and so do β and q by (2.22) and (2.23).

The problem may be reduced to the following boundary value problem:

$$\varphi(t) + t\overline{\varphi'(t)} + \overline{\psi(t)} = f(t) + C(t), \quad t \in L, \tag{2.35}$$

where $C(t)$ represents undetermined constants on different boundary contours and (we have assumed $z = 0 \in S^-$)

$$f(t) = -i\int_0^t (X_n + iY_n)ds, \text{①} \quad t \in L,$$

where s is the arc-length parameter on each boundary contour.

Substituting (2.24) and (2.25) into (2.35), we obtain

$$\varphi_0(t) + (t - \bar{t})\overline{\varphi_0'(t)} + \overline{\psi_0(t)} = f_0(t) + C(t), \tag{2.36}$$

where we have put

$$\begin{aligned} f_0(t) = & \frac{1}{2\pi(\kappa+1)}\sum_{k=1}^{n}(X_k + iY_k)\log\sin\frac{t - z_k}{a} \\ & - \frac{\kappa}{2\pi(\kappa+1)}\sum_{k=1}^{n}(X_k + iY_k)\overline{\log\sin\frac{t - z_k}{a}} \\ & + \frac{t - \bar{t}}{2a\pi(\kappa+1)}\sum_{k=1}^{n}(X_k - iY_k)\overline{\cot\frac{t - z_k}{a}} \\ & - (\beta + \bar{\beta})t - (\kappa\beta - \bar{\beta} + q)\bar{t} + f(t), \end{aligned} \tag{2.37}$$

which is single-valued on L obviously. As for $C(t)$, the constants that it takes on $L_k^{(j)}$'s are not all independent and the relations among them could be determined by the continuous branches of the logarithms chosen in (2.37). For in-

① See Lu [6]. The minus sign occurs in the right-hand member because the elastic body occupies the right (negative) side of the positive sense of L.

stance, let us first choose a definite branch of logsin $\frac{t - z_k}{a}$ on l_k and regard

$$\text{logsin}\,\frac{t + ja\pi - z_k}{a} = \text{logsin}\,\frac{t - z_k}{a} + j\pi i$$

when $t + ja\pi \in l_k^{(j)}$, then it is easily seen, by (2.27),

$$f_0(t + a\pi) - f_0(t) = -\frac{Y - iX}{2} - [(\kappa + 1)\beta + q]a\pi$$

and so

$$C(t + a\pi) - C(t) = \frac{Y - iX}{2} + [(\kappa + 1)\beta + q]a\pi = a\pi(\sigma_- - i\tau_- + 2q).$$

In fact, the number of undetermined constants in (2.35) is actually n: one corresponding to each l_k ($k = 1, \cdots, n$). Moreover, we recall that, when solving (2.36), $\varphi_0(\pm\infty i)$ and $\psi_0(\pm\infty i)$ have been required to be finite.

The above boundary value problem may be reduced to solve a Fredholm integral equation. For the case where there is only one single hole in a periodic strip, it was studied by Mikhlin [1]. For the general case, it may be discussed by a method analogous to the nonperiodic case (see Muskhelishvili [1]).

There is another kind of formulation for the first fundamental problem: besides $X_n + iY_n$, the constants σ_-, τ_- (or σ_+, τ_+) and q being also given, find the equilibrium. The result is the same since $h_\pm$ may be found out at once by (2.22). For example, if the displacements are required also periodic, then $h_\pm$ are readily determined without solving the boundary problem itself.

Second fundamental problem Given the relative displacements on the boundary contours of the periodic holes, the resultant principal vector $X + iY$ of the external stresses along the boundary contours in a periodic strip and the stresses at $z = -\infty i$ (or $z = +\infty i$), find the equilibrium.

Here, the relative displacements mean that they are quasi-periodic but the constant q in (2.13) is not given. As in the first fundamental problem, this is the most general formulation of the problem. It may be reduced to solve a Fredholm integral equation as well.

For convenience of discussion, in the sequel, we always assume that the displacements are also periodic.

4. Stress functions for elastic half-plane

Assume the elastic body occupies the lower half-plane S^-. In this case, the stresses and displacements may be expressed in terms of a single stress function.

When z lies in the upper half-plane S^+, let

$$\bar{\Phi}(z) = \overline{\Phi(\bar{z})},\ \bar{\Psi}(z) = \overline{\Psi(\bar{z})},\ z \in S^+, \tag{2.38}$$

where $\Phi(z)=\varphi'(z)$, $\Psi(z)=\psi'(z)$ are the stress functions defined before and put

$$\Phi(z) = -\bar{\Phi}(z) - z\bar{\Phi}'(z) - \bar{\Psi}(z), \quad z \in S^+. \tag{3.39}$$

Then, (2.1) - (2.3) may be written as

$$\begin{aligned} \sigma_x + \sigma_y &= 2[\Phi(z) + \overline{\Phi(z)}], \\ \sigma_y - \sigma_x + 2i\tau_{xy} &= 2[(\bar{z} - z)\Phi'(z) - \Phi(z) - \bar{\Phi}(z)] \end{aligned} \tag{2.40}$$

or

$$\sigma_y - i\tau_{xy} = \Phi(z) - \Phi(\bar{z}) + (z - \bar{z})\overline{\Phi'(z)}, \tag{2.42}$$

$$2\mu(u' + iv') = \kappa\Phi(z) + \Phi(\bar{z}) - (z - \bar{z})\overline{\Phi'(z)}, \tag{2.43}$$

where
$$u' = \frac{\partial u}{\partial x}, \quad v' = \frac{\partial v}{\partial x},$$

and so

$$2\mu(u + iv) = \kappa\varphi(z) + \varphi(\bar{z}) - (z - \bar{z})\overline{\varphi'(z)} + \text{const}, \tag{2.44}$$

in which we have set $\varphi'(z)=\Phi(z)$ for either $z \in S^-$ or $z \in S^+$. Moreover, on the part of the x-axis without external loads, $\Phi(z)$ is the analytic extension mutually of S^- and S^+.

Now, the stress function $\Phi(z)$ possesses the following property.

Theorem 2.4 *In the periodic problems for isotropic elastic half-plane, $\Phi(z)$ in (2.40) - (2.43) is $a\pi$-periodic for $z \in S^+$ and $z \in S^-$ including its boundary values and bounded at $z = \pm\infty i$.*

Note that, at any point t on the x-axis, where the load is not equal to zero, $\Phi(t)$ is meaningless but $\Phi^+(t) = \Phi^+(t + a\pi)$ and $\Phi^-(t) = \Phi^-(t + a\pi)$ are meaningful in general.

Proof At the first place, assume $z \in S^-$. By (2.40), it is easily verified

$$\Phi'(z + a\pi) = \Phi'(z), \quad z \in S^-.$$

Since S^- is simply connected, it follows immediately

$$\Phi(z + a\pi) = \Phi(z) + C, \quad z \in S^-, \tag{$*$}$$

where C is a constant. By (2.42), the left-hand member of which is periodic by assumption, we have

$$\Phi(\bar{z} + a\pi) = \Phi(\bar{z}) + C, \quad z \in S^-,$$

that is, ($*$) is also valied for $z \in S^+$. We conclude $C = 0$ at once by the periodicity of the left-hand member of (2.43). Hence $\Phi(z)$ is periodic for the whole plane including $\Phi^{\pm}(t)$ for $\mathrm{Im}t = 0$.

The boundedness of $\Phi(z)$ and $\Psi(z)$ at $z = -\infty i$ is already known in § 1.1 and so $\Phi(z)$ is also bounded at $z = +\infty i$ by (2.39). The theorem is proved.

§ 2. Periodic Welding Problems

In the present section, we would solve periodic *welding problems* by using the results of periodic Riemann boundary value problems.

1. The case of uniform material welded

Assume there is a row of periodic holes in the isotropic elastic plane with boundary contours L_0, $L_{\pm 1}$, $L_{\pm 2}$, $\cdots$, oriented counter-clockwisely (Fig. 1.1). Let each hole be welded with a plate of the material same as the elastic plane. Assume the *difference function* between each boundary hole and perimeter of the plate is given:

$$(u^+ + iv^+) - (u^- + iv^-) = g(t), \quad t \in L, \tag{2.45}$$

where $g(t)$ is $a\pi$-periodic with $g'(t) \in H$. Also the stresses σ, τ at $z = \pm\infty i$ are known. Find the elastic equilibrium.

By Corollary 2.1, $\varphi_0(z)$ and $\psi_0(z)$ are periodic and bounded.

Under the above assumptions, we have the following boundary conditions:

$$\varphi^+(t) + t\overline{\varphi'^+(t)} + \overline{\psi^+(t)} = \varphi^-(t) + \overline{\varphi'^-(t)} + \overline{\psi^-(t)}, \quad t \in L, \tag{2.46}$$

$$\kappa\varphi^+(t) - t\overline{\varphi'^+(t)} - \overline{\psi^+(t)} = \kappa\varphi^-(t) - t\overline{\varphi'^-(t)} - \overline{\psi^-(t)} + 2\mu g(t), \quad t \in L. \tag{2.47}$$

From these two conditions, we see that our problem is reduced to two simplest boundary value problems:

$$\varphi^+(t) - \varphi^-(t) = \frac{2\mu}{\kappa + 1} g(t),$$

$$\psi^+(t) - \psi^-(t) = \frac{2\mu}{\kappa + 1} h(t),$$

where

$$h(t) = -\overline{g(t)} - \bar{t}g'(t),$$

or,

$$\varphi_0^+(t) - \varphi_0^-(t) = \frac{2\mu}{\kappa + 1} g(t), \quad t \in L, \tag{2.48}$$

$$\psi_0^+(t) - \psi_0^-(t) = \frac{2\mu}{\kappa + 1} h_0(t), \quad t \in L, \tag{2.49}$$

where

$$h_0(t) = -\overline{g(t)} + (t - \bar{t})g'(t). \tag{2.50}$$

Evidently, $h_0(t)$ is already $a\pi$-periodic and $\in H$. Thus, our problem is reduced to solve these two periodic boundary value problems in which it is required $\varphi_0(z)$ and $\psi_0(z)$ to be bounded at $z = \pm\infty i$.

The indices of both these problems are $k = 0$. Therefore, their solutions are respectively

$$\varphi_0(z) = \frac{\mu}{(\kappa + 1)a\pi i}\int_{L_0} g(t)\cot\frac{t - z}{a}dt, \tag{2.51}$$

$$\psi_0(z) = \frac{\mu}{(\kappa + 1)a\pi i}\int_{L_0} h_0(t)\cot\frac{t - z}{a}dt. \tag{2.52}$$

Here, the arbitrary constant terms are omitted since they could only give a rigid translation of the whole elastic system.

The final result may be obtained when $\varphi(z)$ and $\psi(z)$ are evaluated by (2. 26) and (2.27) respectively.

Remark 1 The above discussions remain effective in case L to be a row of periodic cracks, in which case the given $g(t)$ must be zero at the tips of L.

Remark 2 If the considered elastic body is a half-plane or an infinite horizontal strip with a row of periodic holes, then the problem may be reduced to the periodic first fundamintal problem in the related region by using the method similar to that discussed by Muskhelishvili [1]. For the periodic first fundamental problem in the half-plane, please also refer to the first paragraph of the next section.

Example 2.1 Consider the special case: a row of periodic circular holes welded by circular disks of the same material.

Let L_0 be the circle of radius r: $|z| = r$ and the circular disks to be welded be of radius $r + \varepsilon$ ($\varepsilon > 0$), a little larger than r. Assume there are no stresses at points $z = \pm\infty i$. Now,

$$\left.\begin{aligned} g(t) &= -\varepsilon e^{i\theta} = -\frac{\varepsilon t}{r}, \\ h_0(t) &= \frac{\varepsilon\bar{t}}{r} - (t - \bar{t})\frac{\varepsilon}{r} = \frac{2\varepsilon r}{t} - \frac{\varepsilon t}{r}, \end{aligned}\right\} \quad t = re^{i\theta} \in L_0,$$

and $\sigma = \tau = 0$. By (2.51) and (2.52), we have

$$\varphi_0(z) = -\frac{\mu\varepsilon}{(\kappa + 1)a\pi i}\int_{L_0} t\cot\frac{t - z}{a}dt = \begin{cases} -\dfrac{2\mu\varepsilon z}{(\kappa + 1)r}, & z \in S_0^+, \\ 0, & z \in S^-; \end{cases}$$

$$\psi_0(z) = \frac{\mu\varepsilon}{(\kappa + 1)a\pi i}\int_{L_0}\left(\frac{2r}{t} - \frac{t}{r}\right)\cot\frac{t - z}{a}dt$$

$$= \begin{cases} \dfrac{2\mu\varepsilon}{\kappa+1}\left(\dfrac{2r}{z} - \dfrac{z}{r} - \dfrac{2r}{a}\cot\dfrac{z}{a}\right), & z \in S_0^+, \\ -\dfrac{4\mu\varepsilon r}{a(\kappa+1)}\cot\dfrac{z}{a}, & z \in S^-. \end{cases}$$

Thus, we obtain finally

$$\varphi(z) = \begin{cases} -\dfrac{2\mu\varepsilon}{\kappa+1}\dfrac{z}{r}, & z \in S_0^+, \\ 0, & z \in S^-, \end{cases}$$

$$\psi(z) = \begin{cases} \dfrac{4\mu\varepsilon r}{\kappa+1}\left(\dfrac{1}{z} - \dfrac{1}{a}\cot\dfrac{z}{a}\right), & z \in S_0^+, \\ -\dfrac{4\mu\varepsilon r}{a(\kappa+1)}\cot\dfrac{z}{a}, & z \in S^-. \end{cases}$$

For $z \in S_j^+ \ (j = \pm 1, \ \pm 2, \ \cdots)$, we need only make periodic extension.

When $a \to +\infty$, the expressions for the infinite plane with a circular hole welded by a circular disk of the same medium are obtained:

$$\varphi(z) = \begin{cases} -\dfrac{2\mu\varepsilon}{\kappa+1}\dfrac{z}{r}, & z \in S^+, \\ 0, & z \in S^-, \end{cases}$$

$$\psi(z) = \begin{cases} 0, & z \in S^+, \\ -\dfrac{4\mu\varepsilon r}{(\kappa+1)z}, & z \in S^-, \end{cases}$$

which are identical to the results in Muskhelishvili [1].

2. The case of welded material with the same shearing modulus

Assume the elastic plane as before and the welded material is different but with the same shearing modulus μ (only its Young's modulus or Poisson ratio is different). In this case, the problem may be reduced to certain Riemann boundary value problem, slightly less simpler.

Assume $\kappa = \kappa^+$ for the plate S^+ and $\kappa = \kappa^-$ for the plane weakened by the periodic holes. The condition of equilibrium (2.46) of the stresses on L and the periodic displacement difference on L remain the same as before. The condition of continuity of welding (2.47) is now changed to

$$\begin{aligned} &\kappa^+ \varphi^+(t) - t\overline{\varphi'^+(t)} - \overline{\psi^+(t)} \\ &= \kappa^- \varphi^-(t) - t\overline{\varphi'^-(t)} - \overline{\psi^-(t)} + 2\mu g(t), \quad t \in L. \end{aligned} \tag{2.53}$$

Adding (2.46) and (2.53) together, we get the boundary condition of the periodic Riemann problem for $\varphi(z)$:

$$\varphi^+(t) = \frac{\kappa^- + 1}{\kappa^+ + 1}\varphi^-(t) + \frac{2\mu}{\kappa^+ + 1}g(t). \tag{2.54}$$

Taking conjugates in (2.46) and substituting (2.54) into it, we have

$$\psi^+(t) - \psi^-(t) = -[\overline{\varphi^+(t)} - \overline{\varphi^-(t)}] + \bar{t}[\varphi'^+(t) - \varphi'^-(t)]. \tag{2.55}$$

Then we take periodic functions $\varphi_0(z)$ and $\psi_0(z)$ in place of $\varphi(z)$ and $\psi(z)$ respectively as follows.

Now, (2.36) and (2.27) remain true for z in S^+ and S^- with the modification that β should be changed respectively to [see (2.28)]

$$\beta^+ = \frac{-\sigma + i\tau}{1+\kappa^+}, \quad \beta^- = \frac{-\sigma + i\tau}{1+\kappa^-}, \tag{2.56}$$

where σ and τ are again the common stresses at $z = \pm\infty i$, by which it is obvious that

$$\beta^+(1+\kappa^+) = \beta^-(1+\kappa^-) = -\sigma + i\tau. \tag{2.57}$$

Now, in place of (2.26) and (2.27), we have respectively in S^+ and S^-,

$$\left.\begin{aligned} \varphi^+(z) &= \varphi_0^+(z) + \beta^+ z, \\ \varphi^-(z) &= \varphi_0^-(z) + \beta^- z; \end{aligned}\right\} \tag{2.58}$$

$$\left.\begin{aligned} \psi^+(z) &= \psi_0^+(z) - z\varphi'^+(z) + \kappa^+\bar{\beta}^+ z, \\ \psi^-(z) &= \psi_0^-(z) - z\varphi'^-(z) + \kappa^-\bar{\beta}^- z. \end{aligned}\right\} \tag{2.59}$$

(Here, $\varphi^\pm(z)$ mean $\varphi(z)$ for $z \in S^\pm$ respectively, etc.)

Then, (2.54) becomes the boundary condition for $\varphi_0(z)$:

$$\varphi_0^+(t) = \frac{\kappa^- + 1}{\kappa^+ + 1}\varphi_0^-(t) + \frac{2\mu}{\kappa^+ + 1}g(t), \tag{2.60}$$

and (2.55) becomes that for $\psi_0(z)$:

$$\begin{aligned} \psi_0^+(t) - \psi_0^-(t) &= -[\overline{\varphi_0^+(t)} - \overline{\varphi_0^-(t)}] + (t - \bar{t})[\varphi_0'^+(t) - \varphi_0'^-(t)] \\ &\quad + 2(t - \bar{t})\mathrm{Re}(\beta^+ - \beta^-) \\ &= -[\overline{\varphi_0^+(t)} - \overline{\varphi_0^-(t)}] + (t - \bar{t})[\varphi_0'^+(t) - \varphi_0'^-(t) \\ &\quad + \frac{(\kappa^+ - \kappa^-)\sigma}{(1+\kappa^+)(1+\kappa^-)}]. \end{aligned} \tag{2.61}$$

By (2.60), we get at once (up to a rigid displacement)

$$\varphi_0(z) = \frac{\mu}{(\kappa^\pm + 1)a\pi i}\int_{L_0} g(t)\cot\frac{t-z}{a}dt, \quad z \in S^\pm. \tag{2.62}$$

And then, by substituting it into (2.61), $\psi_0(z)$ may be obtained. Hence, our problem is thoroughly solved.

If $g(t)$ satisfies the condition

$$\int_{L_0} g(t)\cot\frac{t-z}{a}dt = 0, \quad z \in S^-, \tag{2.63}$$

then $\varphi_0^-(z)=0$ while $\varphi_0^+(z)$ is still given by (2.62). Then, by (2.60), we have

$$\varphi_0^+(t) = \frac{2\mu}{\kappa^+ + 1} g(t).$$

Substituting it into (2.61) and assuming $\sigma = \tau = 0$, we get

$$\psi_0^+(t) - \psi_0^-(t) = \frac{2\mu}{\kappa^+ + 1}[-\overline{g(t)} + (t - \bar{t})g'(t)] = \frac{2\mu}{\kappa^+ + 1} h_0(t), \tag{2.64}$$

which is identical to (2.49) with $\kappa^+ = \kappa$. In the mean time, as $\beta^+ = \beta^- = 0$, (2.58) and (2.59) are respectively the same as (2.26) and (2.27) with $\beta = 0$, that means, $\varphi(z)$ and $\psi(z)$ are the same as before.

Similarly, if $\sigma = \tau = 0$ and $g(t)$ fulfills condition (2.63) for $z \in S^+$, then we may obtain analogous results with $\kappa = \kappa^-$ instead of κ^+.

Thus, we have

Theorem 2.5 *Let the periodic isotropic elastic plane and the welded plates have the same shearing modulus μ but different κ, and the external stresses at $z = \pm\infty i$ be zero. If the periodic displacement difference $g(t)$ between the boundaries of the holes and the plates satisfies the condition (2.63) for $z \in S^-$ (or S^+), then the stress distribution in S^+ and S^- is the same as that for the case where the elastic plane and the welded plates have the same $\kappa = \kappa^+$ (or κ^-).*

Remark The remarks 1 and 2 in § 2.1 are also in effect here.

Example 2.2 As in Example 2.1, but κ^+ and κ^- are different. Now $\sigma = \tau = 0$. Since $g(t) = -\varepsilon t/r$ and

$$\int_{L_0} t \cot\frac{t - z}{a} dt = 0, \quad z \in S^-,$$

then, by Theorem 2.5, we immediately know that the results in that example remain valid but with κ replaced by κ^+, i.e.,

$$\varphi(z) = \begin{cases} \dfrac{-2\mu\varepsilon z}{(\kappa^+ + 1)r}, & z \in S_0^+, \\ 0, \quad z \in S^-; \end{cases}$$

$$\psi(z) = \begin{cases} \dfrac{4\mu\varepsilon r}{(\kappa^+ + 1)z}\left(1 - \cot\dfrac{z}{a}\right), & z \in S_0^+, \\ -\dfrac{4\mu\varepsilon r}{(\kappa^+ + 1)a}\cot\dfrac{z}{a}, & z \in S^-. \end{cases}$$

§ 3. Periodic Fundamental Problems of Elastic Half-plane

1. The first fundamental problem

Let the isotropic elastic body occupy the lower half-plane S^- in the z-plane. On the x-axis, denote $z = t$ (t is real). Given the external stresses on the x-axis:

$$\sigma_y(t) = -P(t), \quad \tau_{xy}(t) = T(t), \tag{2.65}$$

being arc-wisely $\in H$ and with period $a\pi$. Here P means the normal pressure distribution. Assume again both the stresses and displacements are periodic and the stresses at $z = -\infty i$ are bounded. Find the stress distribution (and the displacements) on the whole elastic body, called the *periodic first fundamental problem* of the half-plane.

We would prove that

Theorem 2.6 *Under the above assumptions, the solution of the periodic first fundamental problem uniquely exists.*

Proof By (2.42), this problem require us to solve the following periodic boundary value problem:

$$\Phi^+(t) - \Phi^-(t) = P(t) + iT(t), \quad t \in L, \tag{2.66}$$

with $\Phi(\pm\infty i)$ bounded, where L is the x-axis.

As before, we obtain

$$\Phi(z) = \frac{1}{2a\pi i}\int_{L_0} [P(t) + iT(t)]\cot\frac{t-z}{a} + C, \tag{$*$}$$

where C is an undetermined constant and L_0 is the line-seqment $-\frac{1}{2}a\pi \leqslant t \leqslant \frac{1}{2}a\pi$ on the x-axis.

The constant C should be determined by the periodic assumption of the displacements. For this purpose, integrating both sides of ($*$), we have, up to a rigid translation of the elastic body,

$$\varphi(z) = -\frac{1}{2\pi i}\int_{L_0} [P(t) + iT(t)]\log\sin\frac{t-z}{a}dt + Cz,$$

where the logarithm, as a function of t on L_0, may be taken as any definite continuous branch.

Note that, when z increases by $a\pi$ along the horizontal direction, $Z = \frac{t-z}{a}$ decreases by π and so $w = \sin Z$ would change to $-w$. When z lies in the lower (upper) half-plane, arg w increases (decreases) by π. Thus, when $z \in S^-$,

$$\varphi(z + a\pi) = -\frac{1}{2\pi i}\int_{L_0}[P(t) + iT(t)][\log\sin\frac{t - z}{a} + i\pi]dt + C(z + a\pi)$$
$$= \varphi(z) - \frac{1}{2}\int_{L_0}[P(t) + iT(t)]dt + a\pi C,$$
$$\varphi(\bar{z} + a\pi) = -\frac{1}{2\pi i}\int_{L_0}[P(t) + iT(t)][\log\sin\frac{t - z}{a} - i\pi]dt + C(\bar{z} + a\pi)$$
$$= \varphi(\bar{z}) + \frac{1}{2}\int_{L_0}[P(t) + iT(t)]dt + a\pi C.$$

Therefore, by (2.44), on account of periodicity of the displacements,

$$\kappa[\varphi(z + a\pi) - \varphi(z)] + [\varphi(\bar{z} + a\pi) - \varphi(\bar{z})]$$
$$= -\frac{\kappa - 1}{2}\int_{L_0}[P(t) + iT(t)]dt + (\kappa + 1)a\pi C = 0$$

and so

$$C = \frac{\kappa - 1}{\kappa + 1}\frac{1}{2a\pi}\int_{L_0}[P(t) + iT(t)]dt = \frac{1}{2}\frac{\kappa - 1}{\kappa + 1}(P^* + iT^*), \tag{2.67}$$

where we have put

$$P^* = \frac{1}{a\pi}\int_{L_0}P(t)dt, \quad T^* = \frac{1}{a\pi}\int_{L_0}T(t)dt. \tag{2.68}$$

Hence, by substituting the value of C into (*), the unique solution of the problem is obtained:

$$\Phi(z) = \frac{1}{2a\pi i}\int_{L_0}[P(t) + iT(t)]\cot\frac{t - z}{a}dt + \frac{1}{2}\frac{\kappa - 1}{\kappa + 1}(P^* + iT^*). \tag{2.69}$$

The theorem is proved.

Since a rigid rotation of the whole elastic body may be disregarded, the pure imaginary term in the expression of $\Phi(z)$ may be omitted and then

$$\Phi(z) = \frac{1}{2a\pi i}\int_{L_0}[P(t) + iT(t)]\cot\frac{t - z}{a}dt + \frac{1}{2}\frac{\kappa - 1}{\kappa + 1}P^*. \tag{2.70}$$

P^* and T^* have clear mechanical meaning. They are actually the external pressure force and shearing force at $z = -\infty i$ respectively, which may be seen by (2.42). In fact, letting $z \to -\infty i$ in (2.4) and (2.42), and noting that

$$\lim_{z \to -\infty i}(z - \bar{z})\Phi'(z) = 0, \tag{2.71}$$

we get at once

$$\sigma_x(-\infty i) + \sigma_y(-\infty i) = 4\mathrm{Re}\Phi(-\infty i) = -\frac{4}{\kappa+1}P^*,$$
$$\sigma_y(-\infty i) - i\tau_{xy}(-\infty i) = -(P^* + iT^*),$$

and so

$$\sigma_y(-\infty i) = P^*, \ \tau_{xy}(-\infty i) = T^*, \tag{2.72}$$
$$\sigma_x(-\infty i) = -\frac{3-\kappa}{1+\kappa}P^*. \tag{2.73}$$

The meaning of (2.72) is clear in mechanics: taking a periodic half-strip out of S^- and noting the periodicity of stresses, it follows that they cancel each other on the two vertical boundary half-rays, and we immediately obtain the above result.

Reminding that

$$\mu > 0, \ 1 < \kappa < 3, \tag{2.74}$$

we conclude that $\sigma_x(-\infty i) < 0$ by (2.73) when $P^* > 0$.

Thus, we have the following corollary.

Corollary 2.2 *For the periodic first fundamental problem of isotropic elastic half-plane, if the resultant of the external normal stresses on the boundary of a period is a pressure force:*

$$\int_{L_0} P(t)dt = a\pi P^* > 0,$$

then σ_x is a compression at $z = -\infty i$, given by (2.73).

Remark If the periodic condition of the horizontal displacements in the above problem is replaced by quasi-periodicity, i.e., $u(z + a\pi) = u(z) + q$ is assumed, where q is a constant not preassigned, and $\sigma_x(-\infty i)$ is already given, then the problem is uniquely solvable as well.

Example 2.3 Assume periodic partial segments on the boundary of an isotropic elastic half-plane are subjected to periodic uniform pressure. Find the equilibrium (Fig.2.2).

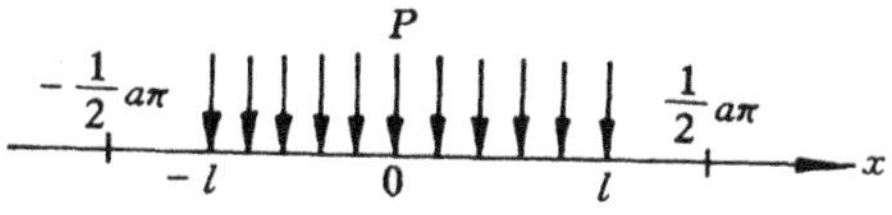

Fig.2.2

That is, on the boundary of a period,

$$P(t) = \begin{cases} P, & |t| \leqslant l, \\ 0, & l < |t| \leqslant \frac{1}{2}a\pi; \end{cases} \qquad T(t) = 0,$$

where P is a given (positive) constant.

Denote the segment $-l \leqslant t \leqslant l$ by γ_0, oriented from left to right. In this case,

$$P^* = \frac{2lP}{a\pi},$$

and so, by (2.70),

$$\Phi(z) = \frac{P}{2a\pi i}\int_{\gamma_0} \cot\frac{t-z}{a}dt + \frac{\kappa-1}{\kappa+1}\frac{lP}{a\pi}. \tag{2.75}$$

By substitution into (2.36) and (2.37), it is readily found that

$$\left.\begin{aligned} \sigma_x + \sigma_y &= \frac{2P}{\pi}\left[\arg\sin\frac{t-z}{a}\right]_{\gamma_0} + \frac{\kappa-1}{\kappa+1}\frac{4lP}{a\pi}, \\ \sigma_y - \sigma_x + 2i\tau_{xy} &= \frac{P(z-\bar{z})\sin\frac{2l}{a}}{a\pi i \sin\frac{l+z}{a}\sin\frac{l-z}{a}} - \frac{\kappa-1}{\kappa+1}\frac{4lP}{a\pi}, \end{aligned}\right\} \tag{2.76}$$

where $\left[\arg\sin\frac{t-z}{a}\right]_{\gamma_0}$ denotes the increment of the argument of $\sin\frac{t-z}{a}$ when t describes γ_0 from $-l$ to $+l$.

If, in (2.75), let $l \to 0$ and $P \to +\infty$ but keep $2lP = P_0$, then it becomes the solution of the corresponding problem when a periodic concentrated pressure P_0 is subjected on the boundary:

$$\Phi(z) = -\frac{P_0}{2a\pi i}\cot\frac{z}{a} + \frac{\kappa-1}{\kappa+1}\frac{P_0}{2a\pi}, \tag{2.77}$$

and the stress formulas (2.76) becomes

$$\left.\begin{aligned} \sigma_x + \sigma_y &= -\frac{2P_0}{a\pi}\operatorname{Im}\cot\frac{z}{a} + \frac{\kappa-1}{\kappa+1}\frac{2P_0}{a\pi}, \\ \sigma_y - \sigma_x + 2i\tau_{xy} &= \frac{P_0(\bar{z}-z)}{a^2\pi i\sin^2\frac{z}{a}} - \frac{\kappa-1}{\kappa+1}\frac{2P_0}{a\pi}. \end{aligned}\right\} \tag{2.78}$$

If let $a \to +\infty$ in (2.75), then we return to the stress distribution when a uniform pressure P is subjected to an interral $\gamma_0[-l, l]$ of the boundary of the half-plane

$$\sigma_x + \sigma_y = \frac{2P}{\pi}[\arg(t-z)]_{\gamma_0},$$

$$\sigma_y - \sigma_x + 2i\tau_{xy} = \frac{2lP(z-\bar{z})}{\pi i(l^2 - z^2)},$$

which is identical to the result in Muskhelishvili [1].

Similarly, if let $a \to +\infty$ in (2.78), we obtain the solution when a concentric pressure P_0 at the origin applied to the boundary of the half-plane:

$$\sigma_x + \sigma_y = -\frac{2P_0}{\pi}\mathrm{Im}\frac{1}{z},$$

$$\sigma_y - \sigma_x + 2i\tau_{xy} = \frac{P_0(\bar{z}-z)}{\pi i z^2}.$$

Example 2.4 The same as Example 2.3, but a uniform shearing stress T is subjected to γ (γ_0 and its periodic congruent segments). Now

$$P(t) = 0,\quad T(t) = \begin{cases} T, & |t| \leqslant l, \\ 0, & l \leqslant |t| \leqslant \frac{1}{2}a\pi. \end{cases}$$

Since $P^* = 0$ in this case, we have, by (2.70),

$$\Phi(z) = \frac{T}{2a\pi}\int_{\gamma_0} \cot\frac{t-z}{a}dt. \tag{2.79}$$

By substitution into (2.36) and (2.38), it is easily found that

$$\left.\begin{aligned} \sigma_x + \sigma_y &= \frac{2T}{\pi}\ln\left|\frac{\sin\frac{l-z}{a}}{\sin\frac{l+z}{a}}\right|, \\ \sigma_y + i\tau_{xy} &= \frac{Ti}{\pi}\left[\operatorname{argsin}\frac{t-z}{a}\right]_{\gamma_0} + \frac{T(z-\bar{z})\sin\frac{2l}{a}}{2a\pi\sin\frac{l+z}{a}\sin\frac{l-z}{a}}. \end{aligned}\right\} \tag{2.80}$$

If, in (2.79) and (2.80), let $l \to 0$ and $T \to +\infty$ but keep $2lT = T_0$ is a constant, then, by discussion analogous to (2.76), the solution of the case of periodic concentrated shearing forces at $z = ak\pi$ ($k = 0, \pm 1, \pm 2, \cdots$) on the boundary is obtained:

$$\Phi(z) = -\frac{T_0}{2a\pi}\cot\frac{z}{a}, \tag{2.81}$$

$$\left.\begin{aligned} \sigma_x + \sigma_y &= -\frac{2T_0}{a\pi}\mathrm{Re}\cot\frac{z}{a}, \\ \sigma_y + i\tau_{xy} &= \frac{T_0 i}{a\pi}\mathrm{Im}\cot\frac{z}{a} + \frac{T_0(\bar{z}-z)}{2a^2\pi\sin^2\frac{z}{a}}. \end{aligned}\right\} \tag{2.82}$$

If let $a \to +\infty$ in (2.80) and (2.82), we may obtain the solutions for the cases of uniform shearing stress on γ_0 and concentric shearing force at origin respectively, which would not be written here.

2. The second fundamental problem

Assume that, on the boundary L of the isotropic elastic half plane S^-, the complex displacement $u^- + iv^- = g(t)$ is given, where $g(t)$ is continuous and $a\pi$-periodic (up to an arbitrary constant term, corresponding to a rigid translation of the whole elastic body), $g'(t) \in H$ arc-wisely, and the principal vector $X + iY$ of the external stresses on a period of the boundary is also given. Again we assume the stresses and the displacements are periodic and the stresses at $z = -\infty i$ are bounded. Find the elastic equilibrium. This is the *periodic second fundamental problem* of the half-plane.

Theorem 2.7 *Under the above assumptions, the solution of the periodic second fundamental problem of the half-plane uniquely exists.*

Proof By (2.39), we want to solve the periodic Riemann boundary value problem

$$\Phi^+(t) + \kappa\Phi^-(t) = 2\mu g'(t), \quad t \in L, \tag{2.83}$$

with $\Phi(\pm\infty i)$ bounded. Its general solution is

$$\Phi(z) = \begin{cases} \Phi^+(z) = \dfrac{\mu}{a\pi i}\displaystyle\int_{L_0} g'(t)\cot\frac{t-z}{a}dt - \kappa C, & z \in S^+, \\ \Phi^-(z) = -\dfrac{\mu}{\kappa a\pi i}\displaystyle\int_{L_0} g'(t)\cot\frac{t-z}{a}dt + C, & z \in S^-, \end{cases} \tag{2.84}$$

where C is an arbitrary constant.

Then, we have, up to a rigid translation,

$$\varphi(z) = \begin{cases} \varphi^+(z) = -\dfrac{\mu}{\pi i}\displaystyle\int_{L_0} g'(t)\log\sin\frac{t-z}{a}dt - \kappa Cz, & z \in S^+, \\ \varphi^-(z) = \dfrac{\mu}{\kappa\pi i}\displaystyle\int_{L_0} g'(t)\log\sin\frac{t-z}{a}dt + Cz, & z \in S^-, \end{cases} \tag{2.85}$$

where the logarithm may be taken as any branch which gives no influence on $\varphi(z)$ since

$$\int_{L_0} g'(t)dt = [g(t)]_{L_0} = 0. \tag{$*$}$$

Since, when $z \in S^-$,

$$\varphi(\bar{z}) = -\frac{\mu}{\pi i}\int_{L_0} g'(t)\log\sin\frac{t-\bar{z}}{a}dt - \kappa C\bar{z}$$

$$= -\frac{\mu}{\pi i}\int_{L_0} g'(t)\,\overline{\log\sin\frac{t-z}{a}}dt - \kappa C\bar{z},$$

by substitution into (2.44), we get

$$2\mu(u+iv) = \frac{2\mu}{\pi}\int_{L_0} g'(t)\arg\sin\frac{t-z}{a}dt$$
$$+ \kappa C(z-\bar{z}) - (z-\bar{z})\overline{\Phi(z)},$$

from which we see that not only $u + iv$ is single-valued but also is periodic as the first term on its right side is invariant when z is changed to $z + a\pi$ on account of (*).

Consider the stresses at $z = -\infty i$ so as to determine C. According to the condition of equilibrium, we have

$$\sigma_y(-\infty i) = \frac{Y}{a\pi},\quad \tau_{xy}(-\infty i) = \frac{X}{a\pi}, \tag{2.86}$$

and, by (2.84),

$$\Phi(+\infty i) = -\kappa C,\quad \Phi(-\infty i) = C.$$

Moreover, (2.71) remains valid.

Therefore, letting $z \to -\infty i$ in (2.42) and noting (2.86), we have

$$C = \frac{Y - iX}{(\kappa+1)a\pi}.$$

Thus, we obtain at length

$$\Phi(z) = \begin{cases} \dfrac{\mu}{a\pi i}\displaystyle\int_{L_0} g'(t)\cot\frac{t-z}{a}dt - \dfrac{\kappa(Y-iX)}{(\kappa+1)a\pi}, & z \in S^+, \\ -\dfrac{\mu}{\kappa a\pi i}\displaystyle\int_{L_0} g'(t)\cot\frac{t-z}{a}dt + \dfrac{Y-iX}{(\kappa+1)a\pi}, & z \in S^-. \end{cases} \tag{2.87}$$

The theorem is proved.

The following two corollaries are evident.

Corollary 2.3 $\sigma_x(-\infty i) = 0$ *iff* $Y = 0$.

Corollary 2.4 *The horizontal displacement u is bounded when and only when $X = 0$ and so does the vertical displacement v when and only when $Y = 0$.*

Example 2.5 Assume that the periodic displacement is of wedge type on the boundary of the half-plane, i.e., the displacement on L_0 is given by

$$g(t) = \begin{cases} \varepsilon\left(\dfrac{|t|}{l} - 1\right)i, & |t| \leqslant l, \\ 0, & l < |t| \leqslant \dfrac{1}{2}a\pi, \end{cases}$$

and the principal vector of the external stresses on L_0 is zero. Find the elastic equilibrium (Fig.2.3).

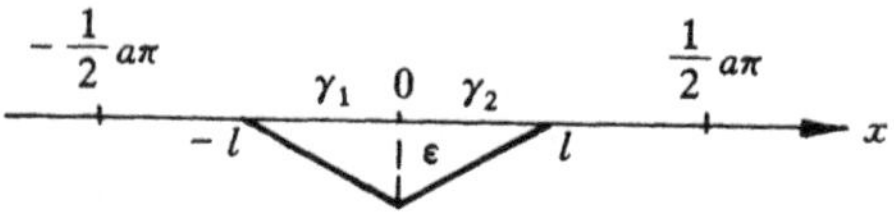

Fig.2.3

In this case, $C=0$. Therefore, when $z\in S^-$, by (2.87),

$$\Phi(z) = \frac{\mu\varepsilon}{\kappa l a\pi}\left(\int_{\gamma_1}\cot\frac{t-z}{a}dt - \int_{\gamma_2}\cot\frac{t-z}{a}dt\right)$$

$$= \frac{\mu\varepsilon}{\kappa l\pi}\left\{\ln\left|\frac{\sin^2\frac{z}{a}}{\sin\frac{l+z}{a}\sin\frac{l-z}{a}}\right| + i\left[\arg\sin\frac{t-z}{a}\right]_{\gamma_1} - i\left[\arg\sin\frac{t-z}{a}\right]_{\gamma_2}\right\},$$

where γ_1 and γ_2 denote the directed segments from $-l$ to 0 and 0 to $+l$ respectively, and, by (2.87), for $z\in S^-$,

$$\Phi(\bar{z}) = \kappa\overline{\Phi(z)} \quad \text{or} \quad \bar{\Phi}(z) = \kappa\Phi(z).$$

Then, by substitution into (2.36) and (2.37), the formulas of stress distribution are obtained:

$$\sigma_x + \sigma_y = 4\mathrm{Re}\Phi(z) = \frac{4\mu\varepsilon}{\kappa l\pi}\ln\left|\frac{\sin^2\frac{z}{a}}{\sin\frac{l+z}{a}\sin\frac{l-z}{a}}\right|,$$

$$\sigma_y - \sigma_x + 2i\tau_{xy} = -2(\kappa+1)\Phi(z) - 2(z-\bar{z})\Phi'(z)$$

$$= -\frac{2\mu\varepsilon(\kappa+1)}{\kappa l\pi}\left\{\ln\left|\frac{\sin^2\frac{z}{a}}{\sin\frac{l+z}{a}\sin\frac{l-z}{a}}\right| + i\left[\arg\sin\frac{t-z}{a}\right]_{\gamma_1} - i\left[\arg\sin\frac{t-z}{a}\right]_{\gamma_2}\right\}$$

$$-\frac{2\mu\varepsilon(z-\bar{z})}{\kappa l\pi a}\left[2\cot\frac{z}{a} + \frac{\sin\frac{2z}{a}}{\sin\frac{l+z}{a}\sin\frac{l-z}{a}}\right].$$

If let $a\to+\infty$, then we obtain the formulas of stress distribution for the case of boundary displacement of single wedge type:

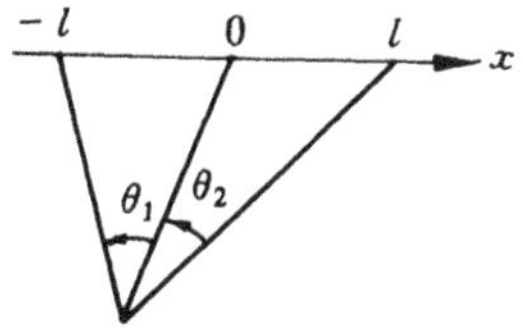

Fig.2.4

$$\sigma_x + \sigma_y = \frac{4\mu\varepsilon}{\kappa l\pi}\ln\left|\frac{z^2}{l^2 - z^2}\right|,$$

$$\sigma_y - \sigma_x + 2i\tau_{xy} = -\frac{2\mu\varepsilon(\kappa + 1)}{\kappa\pi l}\left\{\ln\left|\frac{z^2}{l^2 - z^2}\right| + i(\theta_2 - \theta_1)\right\} - \frac{4\mu\varepsilon(z - \bar{z})}{\kappa l\pi}\left(\frac{1}{z} + \frac{z}{l^2 - z^2}\right), \quad z \in S^-,$$

where θ_1 and θ_2 are positive angles ($<\pi$) as shown in Fig.2.4.

3. The mixed fundamental problem

Let the elastic half-plane S^- as before. On the boundary L_0 in a period, given displacement (up to a constant term) on its sub-segment $\gamma_0(-l \leqslant t \leqslant l)$:

$$u^- + iv^- = g(t), \quad t \in \gamma_0,$$

where $g(t)$ is continuous with $g'(t) \in H$, and on $\gamma_0' = L_0 - \gamma_0$, given the external stresses, for briefness, to be zero:

$$\sigma_y^-(t) = \tau_{xy}^-(t) = 0, \quad l \leqslant |t| \leqslant \frac{1}{2}a\pi.$$

Besides, the principal vector $X + iY$ of the external stresses on γ_0 is also given. All the conditions given are $a\pi$-periodic. Again assume the stresses and the displacements are periodic and the stresses at $z = -\infty i$ are bounded. Find the elastic equilibrium. The problem is called the *periodic mixed fundamental problem*.

Thereem 2.8 *Under the above conditions, the solution of the periodic mixed fundamental problem uniquely exists.*

Proof In general, the stresses near $z = \pm l$ may be unbounded.

Similar to the reasoning in the previous paragraph, now we should solve the periodic boundary value problem

$$\Phi^+(t) + \kappa\Phi^-(t) = 2\mu g'(t), \quad t \in \gamma_0. \tag{2.88}$$

This is the same as the particular case discussed in § 2.2 of Chapter I, in which K is here understood by κ. The solution would be looked for in class h_0.

By (1.48), the general solution of this problem is

$$\Phi(z) = \frac{\mu X(z)}{a\pi i}\int_{\gamma_0}\frac{g'(t)}{X^+(t)}\cot\frac{t-z}{a}dt + X(z)(C_0\tan\frac{z}{a} + C_1), \tag{2.89}$$

where the canonical function

$$X(z) = (\tan\frac{z}{a} + \tan\frac{l}{a})^{-\frac{1}{2}+i\beta}(\tan\frac{z}{a} - \tan\frac{l}{a})^{-\frac{1}{2}-i\beta}, \tag{2.90}$$

in which

$$\beta = \frac{\ln\kappa}{2\pi}. \tag{2.91}$$

Here $X(z)$ has been taken as a definite branch such that, e.g.,

$$\lim_{z\to\pm\frac{1}{2}a\pi}\tan\frac{z}{a}X(z) = 1, \tag{2.92}$$

which is the branch described below, as easily proved. Set

$$\zeta = \tan\frac{z}{a}, \quad L = \tan\frac{l}{a},$$

and cut the ζ-plane along the real axis from $-L$ to $+L$. The branch chosen is such one as shown in Fig. 2.5:

$$\arg(\zeta + L) = \theta_1, \quad \arg(\zeta - L) = \theta_2,$$

where $0<\theta_1,\ \theta_2<\pi$.

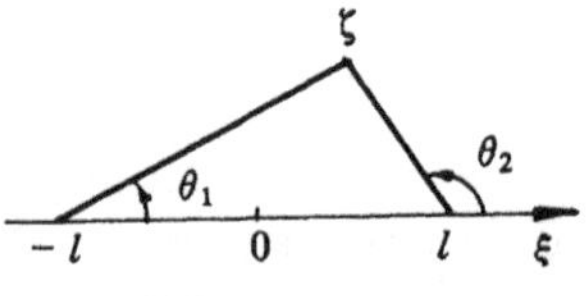

Fig. 2.5

$\Phi(z)$ is already $a\pi$-periodic for any C_0, C_1, and so do the stresses. To determine C_0 and C_1, it is sufficient to use the periodic condition of the displacements and the condition of equilibrium of the stresses at $z = -\infty i$.

In fact, noting that

$$X(-\infty i) = \exp\{(-\frac{1}{2} + i\beta)\log(-i + L) - (\frac{1}{2} + i\beta)\log(-i - L)\}$$

$$= \exp\{(\frac{1}{2} - i\beta)[\ln\cos\frac{l}{a} + i(\frac{\pi}{2} - \frac{l}{a})] + (\frac{1}{2} + i\beta)[\ln\cos\frac{l}{a} + i(\frac{\pi}{2} + \frac{i}{a})]\} = ie^{\frac{-2\beta l}{a}}\cos\frac{l}{a}.$$

Similarly,
$$X(+\infty i) = -ie^{\frac{2\beta l}{a}}\cos\frac{l}{a}.$$

If denote the ratio of the length of γ_0 to the period $a\pi$ (the length of L_0) by

$$\lambda = \frac{2l}{a\pi}$$

and put

$$\left.\begin{aligned} A &= \frac{2\beta l}{a} = \lambda\ln\sqrt{\kappa}, \\ B &= \beta\pi - \frac{2\beta l}{a} = (1-\lambda)\ln\sqrt{\kappa}, \end{aligned}\right\} \qquad (2.93)$$

then

$$X(+\infty i) = -ie^{A}\cos\frac{l}{a}, \quad X(-\infty i) = ie^{-A}\cos\frac{l}{a}. \qquad (2.94)$$

Let Λ_+ and Λ_- be the horizontal segments of a period in the upper and lower half-planes respectively (Fig. 2.6). The integrals $\int_{\Lambda_\pm} X(z)dz$ are evidently independent of the heights of $\Lambda_\pm$.

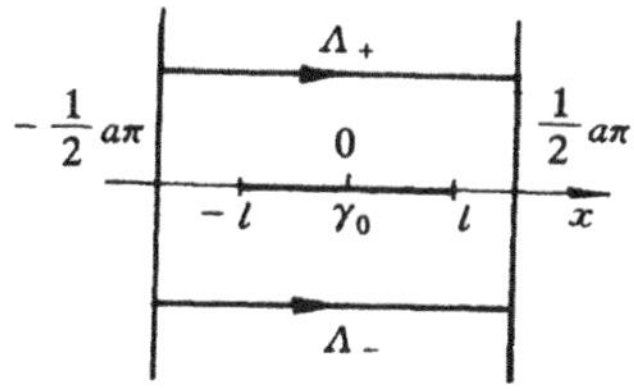

Fig. 2.6

Thus, we have

$$\left.\begin{aligned} \int_{\Lambda_+} X(z)dz &= a\pi X(+\infty i) = -a\pi ie^{A}\cos\frac{l}{a}, \\ \int_{\Lambda_-} X(z)dz &= a\pi X(-\infty i) = a\pi ie^{-A}\cos\frac{l}{a}. \end{aligned}\right\} \qquad (2.95)$$

Similarly,

$$\left.\begin{aligned}\int_{\Lambda_+} X(z)\tan\frac{z}{a}dz &= a\pi e^{A}\cos\frac{l}{a},\\ \int_{\Lambda_-} X(z)\tan\frac{z}{a}dz &= a\pi e^{-A}\cos\frac{l}{a},\end{aligned}\right\} \tag{2.96}$$

Moreover, for $t \in \gamma_0$, we have

$$\left.\begin{aligned}\int_{\Lambda_+} X(z)\cot\frac{t-z}{a}dz &= a\pi e^{A}\cos\frac{l}{a},\\ \int_{\Lambda_-} X(z)\cot\frac{t-z}{a}dz &= a\pi e^{-A}\cos\frac{l}{a}.\end{aligned}\right\} \tag{2.97}$$

Then, by using (2.95) – (2.97), it is easily seen that, for $z \in S^-$,

$$\begin{aligned}\varphi(z+a\pi)-\varphi(z) &= \int_{\Lambda_-}\Phi(z)dz\\ &= a\pi e^{-A}\cos\frac{l}{a}\left\{\frac{\mu}{a\pi i}\int_{\gamma_0}\frac{g'(t)}{X^+(t)}dt + C_0 + iC_1\right\},\\ \varphi(\bar{z}+a\pi)-\varphi(\bar{z}) &= \int_{\Lambda_+}\Phi(z)dz\\ &= a\pi e^{A}\cos\frac{l}{a}\left\{\frac{\mu}{a\pi i}\int_{\gamma_0}\frac{g'(t)}{X^+(t)}dt + C_0 - iC_1\right\}.\end{aligned}$$

Thus, by (2.44), the condition of periodicity of the displacements becomes

$$\begin{aligned}&\kappa e^{-A}\left\{\frac{\mu}{a\pi i}\int_{\gamma_0}\frac{g'(t)}{X^+(t)}dt + C_0 + iC_1\right\}\\ &+ e^{A}\left\{\frac{\mu}{a\pi i}\int_{\gamma_0}\frac{g'(t)}{X^+(t)}dt + C_0 - iC_1\right\} = 0.\end{aligned}$$

After simplification, we have, on account of $e^B = \sqrt{\kappa}e^{-A}$,

$$C_0 + iC_1\tanh B = -\frac{\mu}{a\pi i}\int_{L_0}\frac{g'(t)}{X^+(t)}dt. \tag{2.98}$$

Next, consider the condition of equilibrium of the stresses at $z = -\infty i$:

$$\sigma_y(-\infty i) = \frac{Y}{a\pi},\quad \tau_{xy}(-\infty i) = \frac{X}{a\pi}.$$

Note that

$$\begin{aligned}\Phi(+\infty i) &= -ie^{A}\cos\frac{l}{a}\left\{\frac{\mu}{a\pi}\int_{\gamma_0}\frac{g'(t)}{X^+(t)}dt + C_1 + C_0 i\right\},\\ \Phi(-\infty i) &= ie^{-A}\cos\frac{l}{a}\left\{-\frac{\mu}{a\pi}\int_{\gamma_0}\frac{g'(t)}{X^+(t)}dt + C_1 - C_0 i\right\}\end{aligned}$$

and (2.71) remains valid in this case, we have, by (2.42), the condition of equilibrium becoming

$$\frac{Y - iX}{2a\pi} = i\cos\frac{l}{a}\left\{\frac{\mu\sinh A}{a\pi}\int_{\gamma_0}\frac{g'(t)}{X^+(t)}dt + C_1\cosh A + iC_0\sinh A\right\}, \tag{2.99}$$

i.e.,

$$C_0\tanh A - iC_1 = \frac{-Y + iX}{2a\pi\cos\frac{l}{a}\cosh A} - \frac{\mu\tanh A}{a\pi i}\int_{\gamma_0}\frac{g'(t)}{X^+(t)}dt. \quad (2.99)'$$

Thus, C_0 and C_1 may be uniquely determined by (2.98) and (2.99) since both A and $B > 0$.

Example 2.6 *Periodic stamps with horizontal rectilinear base*

Assume there is a set of periodic stamps pressed on the isotropic elastic half-plane S^-, each of which has a horizontal line segment of length $2l$ as its base. Suppose on each stamp a pressure force P_0 is subjected and there is no external stresses on the boundary out of the stamps. Find the elastic equilibrium.

Here, $g'(t) = 0$ on γ_0 and $X = 0$, $Y = -P$. Now (2.98) and (2.99) become respectively

$$C_0 + iC_1\tanh B = 0,$$
$$C_0\tanh A - iC_1 = \frac{P_0}{2a\pi\cos\frac{l}{a}\cosh A},$$

by which we find

$$C_0 = \frac{\sqrt{\kappa}P_0\sinh B}{(\kappa + 1)a\pi\cos\frac{l}{a}},$$
$$C_1 = \frac{i\sqrt{\kappa}P_0\cosh B}{(\kappa + 1)a\pi\cos\frac{l}{a}}$$

by virtue of

$$\cosh(A + B) = \cosh\beta\pi = \frac{\kappa + 1}{\sqrt{\kappa}}.$$

Substituting into (2.89), we obtain

$$\Phi(z) = \frac{\sqrt{\kappa}P_0X(z)}{(\kappa + 1)a\pi\cos\frac{l}{a}}\left(\sinh B\tan\frac{z}{a} + i\cosh B\right), \tag{2.100}$$

where $X(z)$ is given by (2.90).

Let us evaluate the stress distribution on the boundary right beneath the stamps.

Obviously, $\theta_1 = 0$ and $\theta_2 = \pi$ as z tends to t on γ_0 from the upper half-plane (Fig.2.5). Therefore,

$$X^+(t) = \exp\left\{(-\frac{1}{2} + i\beta)\ln\left|\tan\frac{t}{a} + \tan\frac{l}{a}\right| - (\frac{1}{2} + i\beta)\left[\ln\left|\tan\frac{t}{a} - \tan\frac{l}{a}\right| + \pi i\right]\right\}$$
$$= -\frac{i\sqrt{\kappa}}{\sqrt{\tan^2\frac{l}{a} - \tan^2\frac{t}{a}}}\exp\left\{i\beta\ln\left|\frac{\sin\frac{l+t}{a}}{\sin\frac{l-t}{a}}\right|\right\}. \qquad (2.101)$$

Substituting it into (2.100), we have

$$\Phi^+(t) = \frac{\kappa P_0 \exp\left\{i\beta\ln\left|\frac{\sin\frac{l+t}{a}}{\sin\frac{l-t}{a}}\right|\right\}}{(\kappa+1)a\pi\sqrt{\sin\frac{l+t}{a}\sin\frac{l-t}{a}}}\left(\cosh B\cos\frac{t}{a} - i\sinh B\sin\frac{t}{a}\right).$$

If denote the distribution of the pressure and the shearing stress on γ_0 by $P(t)$ and $T(t)$ respectively, then, by (2.42),

$$P(t) + iT(t) = -\sigma_y(t) + i\tau_{xy}(t) = \Phi^+(t) - \Phi^-(t) = \frac{\kappa+1}{\kappa}\Phi^+(t),$$

and so, finally,

$$P(t) + iT(t) = \frac{P_0}{a\pi}\frac{\cosh B\cos\frac{t}{a} - i\sinh B\sin\frac{t}{a}}{\sqrt{\sin\frac{l+t}{a}\sin\frac{l-t}{a}}}\exp\left\{i\beta\ln\left|\frac{\sin\frac{l+t}{a}}{\sin\frac{l-t}{a}}\right|\right\}. \qquad (2.102)$$

We may get the solution for the case of a single stamp pressed on the x-axis as $a \to +\infty$. At this time, $B \to \ln\sqrt{\kappa}$ by (2.93) and so $\cosh B \to \frac{\kappa+1}{2\sqrt{\kappa}}$, $\sinh B \to \frac{\kappa-1}{2\sqrt{\kappa}}$. Thus, we obtain

$$P(t) + iT(t) = \frac{(\kappa+1)P_0}{2\pi\sqrt{\kappa}\sqrt{l^2 - t^2}}\exp\left\{i\beta\ln\left|\frac{l+t}{l-t}\right|\right\},$$

which is identical to Abramov's formula (See Muskhelishvili [1], § 114a)①.

Example 2.7 *Periodic stamps with inclined rectilinear base*

The bases are as those in the previous example but with a small angle of inclination ε with respect to the boundary of the half-plane.

Assume the principal vector of the external stresses on each stamp is zero: $X = Y = 0$, and there are no external stresses on the boundary out of the stamps.

At this time,

$$g(t) = i\varepsilon t \quad \text{or} \quad g'(t) = i\varepsilon, \ t \in \gamma_0.$$

Now, (2.98) and (2.99) become respectively

$$C_0 + iC_1 \tanh B = -\frac{\mu\varepsilon}{a\pi}\int_{\gamma_0}\frac{dt}{X^+(t)},$$

$$C_0 \tanh A - iC_1 = -\frac{\mu\varepsilon}{a\pi}\tanh A\int_{\gamma_0}\frac{dt}{X^+(t)},$$

which follows at once

$$C_1 = 0, \ C_0 = -\frac{\mu\varepsilon}{a\pi}\int_{\gamma_0}\frac{dt}{X^+(t)}.$$

Substituting into (2.89), we obtain, after simplification,

$$\Phi(z) = \frac{\mu\varepsilon X(z)}{a\pi}\int_{\gamma_0}\left(\cot\frac{t-z}{a} - \tan\frac{z}{a}\right)\frac{dt}{X^+(t)}. \tag{2.103}$$

For further calculations, put

$$I(z) = \int_{\gamma_0}\left(\cot\frac{t-z}{a} - \tan\frac{z}{a}\right)\frac{dt}{X^+(t)}.$$

Using transformation $u = \tan\frac{t}{a}$ and letting $\zeta = \tan\frac{z}{a}$, $L = \tan\frac{l}{a}$, we have

$$\frac{1}{a}\cos^2\frac{z}{a}I(z) = I_*(\zeta) = \int_{-L}^{L}\frac{du}{(1+u^2)(u-\zeta)X_*^+(u)},$$

where

$$X_*(\zeta) = X(z) = (\zeta + L)^{-\frac{1}{2}+i\beta}(\zeta - L)^{-\frac{1}{2}-i\beta}$$

with the branch taken as before, i.e.,

$$\lim_{\zeta\to\infty}\zeta X_*(\zeta) = 1.$$

In order to evaluate $I_*(\zeta)$, introduce an auxiliary function

① In that book, a factor 2 is lost in the denominator on the right side of formula (8).

$$\Omega(\zeta) = \frac{1}{2\pi i}\int_{\Lambda} \frac{dw}{(1 + w^2)X_*(w)(w - \zeta)} = \frac{1}{2\pi i}\int_{\Lambda} \frac{f(w)}{w - \zeta}dw,$$

where

$$f(\zeta) = \frac{1}{(1 + \zeta^2)X_*(\zeta)},$$

in which Λ is an arbitrary smooth closed contour surrounding the segment $-L \leqslant u \leqslant L$ leaving ζ in its exterior and the integral is taken along Λ clockwisely. Note that
$X_*^+(u) = -\kappa X_*^-(u)$. It is easily seen that

$$\Omega(\zeta) = \frac{\kappa + 1}{2\pi i}I_*(\zeta) \quad \text{or} \quad I_*(\zeta) = \frac{2\pi i}{\kappa + 1}\Omega(\zeta). \qquad (*)$$

Let us evaluate $\Omega(\zeta)$. By (2.94), the principal parts of $f(\zeta)$ at $\zeta = i$ and $\zeta = -i$ are respectively

$$\frac{1}{2i(\zeta - i)X_*(i)} = \frac{1}{2i(\zeta - i)X(+\infty i)} = \frac{e^{-A}}{2(\zeta - i)\cos\frac{l}{a}},$$

$$-\frac{1}{2i(\zeta + i)X_*(-i)} = -\frac{1}{2i(\zeta + i)X(-\infty i)} = \frac{e^{A}}{2(\zeta + i)\cos\frac{l}{a}}.$$

Then, by the residue theorem formula, we have

$$\Omega(\zeta) = \frac{1}{(1 + \zeta^2)X_*(\zeta)} - \frac{e^{-A}}{2(\zeta - i)\cos\frac{l}{a}} - \frac{e^{A}}{2(\zeta + i)\cos\frac{l}{a}}$$

$$= \frac{\cos^2\frac{z}{a}}{X(z)} - \frac{\cos^2\frac{z}{a}}{\cos\frac{l}{a}}\left(\cosh A \tan\frac{z}{a} - i\sinh A\right).$$

Substituting it into (*) and returning to $I(z)$, we get

$$I(z) = \frac{2a\pi i}{\kappa + 1}\left\{\frac{1}{X(z)} - \sec\frac{l}{a}\left(\cosh A \tan\frac{z}{a} - i\sinh A\right)\right\}. \qquad (2.104)$$

By (2.103), we have finally

$$\Phi(z) = \frac{2\mu\varepsilon i}{\kappa + 1}\left\{1 - \frac{X(z)}{\cos\frac{l}{a}}\left(\cosh A \tan\frac{z}{a} - i\sinh A\right)\right\}. \qquad (2.105)$$

Thus, our problem is solved.

It is easily known, as in the previons example, right beneath the stamps,

$$P(t) + iT(t) = \Phi^+(t) - \Phi^-(t) = \frac{\kappa+1}{\kappa}\Phi^+(t) - \frac{2\mu\varepsilon i}{\kappa}$$
$$= -\frac{2\mu\varepsilon i X^+(t)}{\kappa\cos\frac{l}{a}}\left(\cosh A\tan\frac{t}{a} - i\sinh A\right),$$

and hence, by (2.101),

$$P(t) + iT(t) = -\frac{2\mu\varepsilon\left(\cosh A\sin\frac{t}{a} - i\sinh A\cos\frac{t}{a}\right)}{\sqrt{\kappa}\sqrt{\sin\frac{l+t}{a}\sin\frac{l-t}{a}}}\exp\left\{i\ln\left|\frac{\sin\frac{l+t}{a}}{\sin\frac{l-t}{a}}\right|\right\}. \tag{2.106}$$

If let $a \to +\infty$ in (2.105) and (2.106), and note that

$$\lim_{a\to+\infty}\frac{X(z)}{a} = X_0(z) = (z+l)^{-\frac{1}{2}+i\beta}(z-l)^{-\frac{1}{2}-i\beta},$$
$$\lim_{a\to+\infty}\cosh A = 1, \quad \lim_{a\to+\infty} a\sinh A = 2l\beta,$$

then the corresponding formulas for the case of a single inclined straight base of the stamp in a period are obtained:

$$\Phi(z) = \frac{2\mu\varepsilon i}{\kappa+1}\{1 - (z - 2l\beta i)X_0(z)\},$$
$$P(t) + iT(t) = -\frac{2\mu\varepsilon i}{\kappa}(t - 2l\beta i)X_0^+(t)$$
$$= -\frac{2\mu\varepsilon}{\sqrt{\kappa}}\frac{t - 2l\beta i}{\sqrt{l^2 - t^2}}\exp\left\{i\beta\ln\left|\frac{l+t}{l-t}\right|\right\},$$

which are identical to the results in Muskhelishvili [1].

§ 4. Periodic Contact Problems

1. The case without friction

Assume a series of periodic stamps (with bases of the same shape) are pressed on the boundary of an isotropic elastic half-plane S^-. In the present paragraph, it is assumed that there exists no friction between the stamps and the half-plane. Out of the stamps, the condition subjected to the boundary of the half-plane are the same as in Example 2.7, i.e., $\sigma_y = \tau_{xy} = 0$. Right beneath the bases of the stamps, only the periodic vertical displacement $v(t)$ is given while the horizontal displacement $u(t)$ is unknown. On the boundary L_0, let the interval pressed be γ_0: $-l \leqslant t \leqslant l$. Since there is no friction, $\tau_{xy} = 0$ also on γ_0 but σ_y is unknown. Besides, assume the load applied to each stamp is a positive pressure force P_0, i.e., $Y = -P_0$ and $X = 0$. Find the elastic equilibrium.

Let $y = f(x)$ be the equation of the base of the stamp pressed on S^-, and $f'(x) \in H$. Thus, the boundary conditions on L_0 are :

$$\tau_{xy}^{-}(t) = 0,\ t \in L_0;\ \sigma_y^{-}(t) = 0,\ t \in L_0 - \gamma_0;$$
$$v^{-}(t) = f(t),\ t \in \gamma_0;$$

and the principal vector of the external stresses on γ_0 is $X + iY = -P_0 i$.

Under these boundary conditions, by the principle of equilibrium, we readily know that

$$\sigma = \sigma_y(-\infty i) = -\frac{P_0}{a\pi},\quad \tau = \tau_{xy}(-\infty i) = 0. \tag{2.107}$$

Note that $\Phi(z)$ is holomorphic in the plane cut by γ (γ_0 and its periodic congruents). By (2.42), obviously,

$$\Phi^{+}(t) - \Phi^{-}(t) = \overline{\Phi}^{-}(t) - \overline{\Phi}^{+}(t),\ t \in \gamma_0,$$

or, what is the same,

$$\Phi^{+}(t) + \overline{\Phi}^{+}(t) = \Phi^{-}(t) + \overline{\Phi}^{-}(t),\ t \in \gamma_0,$$

since $\tau_{xy} = 0$ on γ_0. This means that $\Phi(z) + \overline{\Phi}(z)$ is holomorphic in the entire plane. It is bounded at $z = \pm\infty i$ and periodic. By the Liouville theorem (after a certain conformal mapping), it must be a constant:

$$\Phi(z) + \overline{\Phi}(z) = 2\beta_2. \tag{2.108}$$

When t situates on γ, $\tau_{xy} = 0$. Then it is evident by (2.41) that β_2 is real.

On γ, by (2.43), we have

$$2\mu(u' + iv') = \kappa\Phi^{-}(t) + \Phi^{+}(t).$$

Taking conjugates on both sides and using (2.108), we get

$$2\mu(u' - iv') = -[\kappa\Phi^{+}(t) + \Phi^{-}(t)] + 2(\kappa + 1)\beta_2.$$

Subtracting this equality from the above one and noting that $v' = f'(t)$ on γ_0, we obtain the Riemann boundary value problem

$$\Phi^{+}(t) + \Phi^{-}(t) = \frac{4\mu i}{\kappa + 1} f'(t) + 2\beta_2,\ t \in \gamma_0. \tag{2.109}$$

We want to search for its solution bounded at $z = \pm\infty i$. We may use (1.27)′. But it is obvious that, for the free term $2\beta_2$ (and $f'(t) = 0$), it has a particular solution β_2. Hence, the general solution of (2.109) is

$$\Phi(z) = \frac{2\mu}{(\kappa + 1)a\pi\sqrt{R(z)}}\int_{\gamma_0} f'(t)\sqrt{R(t)}\cot\frac{t - z}{a}dt + \frac{\beta_0\tan\frac{z}{a} + \beta_1}{\sqrt{R(z)}} + \beta_2, \tag{2.110}$$

where we have written $X(z)$ as

$$X(z) = \frac{1}{i\sqrt{R(z)}}, \quad R(z) = \tan^2 \frac{l}{a} - \tan^2 \frac{z}{a}, \tag{2.111}$$

in which $\sqrt{R(z)}$ takes positive value $\sqrt{R(t)}$ when z tends to $t \in \gamma_0$ from S^+. It is also seen that, by (2.110),

$$\lim_{z \to \pm\infty i} (z - \bar{z}) \Phi'(z) = 0 \tag{2.112}$$

remains valid.

Let us prove that, under our assumptions, both β_0 and β_1 must be real. Note first that

$$\overline{X}(z) = X(z). \tag{2.113}$$

In fact, $\overline{X}(z) = \overline{X(\bar{z})}$ by definition, $X(z)$ and $\overline{X}(z)$ represent the same radical and so it is only possible $\overline{X}(z) = \pm X(z)$. But, when compare the left side of (2.109) with (2.88), we find that, the canonical function is the same as there, provided that κ is replaced by 1, or A in (2.93) is changed to 0. Thereby, here we should have $\overline{X}(-\infty i) = \overline{X(+\infty i)} = X(-\infty i)$, that means, the positive sign in the above equality should be taken, i. e., (2.113) is valid. Then, by comparing (2.113) with (2.111), we obtain

$$\overline{\sqrt{R(\bar{z})}} = -\sqrt{R(z)}.$$

Therefore, by (2.110),

$$\overline{\Phi}(z) = -\frac{2\mu}{(\kappa+1)a\pi\sqrt{R(z)}} \int_{\gamma_0} f'(t)\sqrt{R(t)} \cot \frac{t-z}{a} dt - \frac{\bar{\beta}_0 \tan \frac{z}{a} + \bar{\beta}_1}{\sqrt{R(z)}} + \beta_2,$$

since $f'(t)$ is a real function. Adding this equality to (2.110), by (2.108), we know that both β_0 and β_1 are real.

The rest of our problem is to determine β_0, β_1 and β_2.

First, consider the periodic condition of the displacements.

Since

$$\sqrt{R(\pm\infty i)} = \pm 1/\cos \frac{l}{a}, \tag{2.114}$$

we have

$$\int_{\Lambda_\pm} \frac{dz}{\sqrt{R(z)}} = \pm a\pi \cos \frac{l}{a},$$

$$\int_{\Lambda_\pm} \frac{\tan \frac{z}{a}}{\sqrt{R(z)}} dz = \int_{\Lambda_\pm} \frac{\cot \frac{t-z}{a}}{\sqrt{R(z)}} dz = a\pi i \cos \frac{l}{a},$$

where $\Lambda_\pm$ are shown in Fig. 2. 6, and hence, by (2. 112), when z locates in S^-,

$$\varphi(z + a\pi) - \varphi(z) = \int_{\Lambda_-} \Phi(z)dz = \frac{2\mu i \cos\frac{l}{a}}{\kappa + 1}\int_{\gamma_0} f'(t)\sqrt{R(t)}\,dt + a\pi\cos\frac{l}{a}(i\beta_0 - \beta_1) + a\pi\beta_2,$$

$$\varphi(\bar{z} + a\pi) - \varphi(\bar{z}) = \int_{\Lambda_+} \Phi(z)dz = \frac{2\mu i \cos\frac{l}{a}}{\kappa + 1}\int_{\gamma_0} f'(t)\sqrt{R(t)}\,dt + a\pi\cos\frac{l}{a}(i\beta_0 + \beta_1) + a\pi\beta_2.$$

Therefore, by (2.44), the periodic condition of $u + iv$ becomes

$$i(\kappa + 1)\beta_0 - (\kappa - 1)\beta_1 + \frac{\kappa + 1}{\cos\frac{l}{a}}\beta_2 = \frac{2\mu}{a\pi i}\int_{\gamma_0} f'(t)\sqrt{R(t)}\,dt.$$

Thus, we obtain

$$\beta_0 = -\frac{2\mu}{(\kappa + 1)a\pi}\int_{\gamma_0} f'(t)\sqrt{R(t)}\,dt, \tag{2.115}$$

$$(\kappa - 1)\beta_1 - \frac{\kappa + 1}{\cos\frac{l}{a}}\beta_2 = 0. \tag{2.116}$$

Next, consider the condition of equilibrium of the stresses at $z = -\infty i$. By (2.110),

$$\Phi(-\infty i) = \frac{2\mu i\cos\frac{l}{a}}{(\kappa + 1)a\pi}\int_{\gamma_0} f'(t)\sqrt{R(t)}\,dt - \beta_1\cos\frac{l}{a} + \beta_2 + i\beta_0\cos\frac{l}{a},$$

$$\Phi(+\infty i) = \frac{2\mu i\cos\frac{l}{a}}{(\kappa + 1)a\pi}\int_{\gamma_0} f'(t)\sqrt{R(t)}\,dt + \beta_1\cos\frac{l}{a} + \beta_2 + i\beta_0\cos\frac{l}{a}.$$

Therefore, by (2.42), we get, on account of (2.112),

$$\sigma_y(-\infty i) - i\tau_{xy}(-\infty i) = \Phi(-\infty i) - \Phi(+\infty i) = -2\beta_1\cos\frac{l}{a},$$

and then, by using (2.107),

$$\beta_1 = \frac{P_0}{2a\pi\cos\frac{l}{a}}. \tag{2.117}$$

Substituting it into (2.116), we get

$$\beta_2 = \frac{\kappa - 1}{\kappa + 1}\frac{P_0}{2a\pi}. \tag{2.118}$$

By substituting all these equalities into (2.110), the required unique solution is finally obtained:

$$\Phi(z) = \frac{2\mu}{(\kappa+1)a\pi\sqrt{R(z)}}\int_{\gamma_0} f'(t)\sqrt{R(t)}\left(\cot\frac{t-z}{a} - \tan\frac{z}{a}\right)dt + \frac{P_0}{2a\pi\cos\frac{l}{a}\sqrt{R(z)}} + \frac{\kappa-1}{\kappa+1}\frac{P_0}{2a\pi}. \tag{2.119}$$

The pressure distribution $P(t)$ right beneath the bases of the stamps could be easily evaluated as follows.

Let $t_0 \in \gamma_0$. Since $\tau_{xy}^-(t_0) = 0$, we have

$$P(t_0) = -\sigma_y^-(t_0) = \Phi^+(t_0) - \Phi^-(t_0).$$

By the generalized Plemelj formula (See § 1.4, Chapter I), it is easily verified that

$$P(t_0) = \frac{4\mu}{(\kappa+1)a\pi\sqrt{R(t_0)}}\int_{-l}^{l} f'(t)\sqrt{R(t)}\left(\cot\frac{t-t_0}{a} - \tan\frac{t_0}{a}\right)dt + \frac{P_0}{a\pi\cos\frac{l}{a}\sqrt{R(t_0)}}. \tag{2.120}$$

Let us consider the case where P_0 is sufficiently small so as the corners of each stamp could not be in contact with the elastic half-plane.

For simplicity, assume the base of each stamp is symmetric, i.e., $f(t)$ is an even function in t belonging to the base passing through the origin. Let the contact segment of this base be γ_0: $-l \leqslant t \leqslant l$, where l is an unknown positive number (l is less than the half width of the stamp). The condition for its determination is $P(\pm l) = 0$. Note that in this case the term $\tan\frac{z}{a}$ in (2.119) may be omitted and then (2.120) becomes

$$\begin{aligned} P(t_0) &= \frac{4\mu}{(\kappa+1)a\pi\sqrt{R(t_0)}}\int_{-l}^{l} f'(t)\sqrt{R(t)}\cot\frac{t-t_0}{a}dt + \frac{P_0}{a\pi\cos\frac{l}{a}\sqrt{R(t_0)}} \\ &= \frac{4\mu}{(\kappa+1)a\pi\sqrt{R(t_0)}}\int_{-l}^{l} \frac{f'(t)[R(t)-R(t_0)]}{\sqrt{R(t)}}\cot\frac{t-t_0}{a}dt \\ &\quad + \frac{4\mu\sqrt{R(t_0)}}{(\kappa+1)a\pi}\int_{-l}^{l}\frac{f'(t)}{\sqrt{R(t)}}\cot\frac{t-t_0}{a}dt + \frac{P_0}{a\pi\cos\frac{l}{a}\sqrt{R(t_0)}}. \end{aligned}$$

But since

$$[R(t) - R(t_0)]\cot\frac{t-t_0}{a} = -\tan\frac{t}{a}\sec^2\frac{t_0}{a} - \tan\frac{t_0}{a}\sec^2\frac{t}{a}$$

and $f'(t)$ is odd, consequently we have

$$P(t_0) = -\frac{4\mu\sec^2\frac{t_0}{a}}{(\kappa+1)a\pi\sqrt{R(t_0)}}\int_{-l}^{l}\frac{f'(t)\tan\frac{t}{a}}{\sqrt{R(t)}}dt + \frac{4\mu\sqrt{R(t_0)}}{(\kappa+1)a\pi}\int_{-l}^{l}\frac{f'(t)}{\sqrt{R(t)}}\cot\frac{t-t_0}{a}dt + \frac{P_0}{a\pi\cos\frac{l}{a}\sqrt{R(t_0)}}$$

$$= -\frac{4\mu\sec^2\frac{t_0}{a}}{(\kappa+1)a\pi\sqrt{R(t_0)}}\int_{-l}^{l}\frac{f'(t)\tan\frac{t}{a}}{\sqrt{R(t)}}dt + \frac{P_0}{a\pi\cos\frac{l}{a}\sqrt{R(t_0)}} + \frac{4\mu\sqrt{R(t_0)}}{(\kappa+1)a\pi}\int_{-l}^{l}\frac{f'(t)\tan\frac{t}{a}}{\sqrt{R(t)}}dt + \frac{4\mu\sqrt{R(t_0)}}{(\kappa+1)a\pi}\int_{-l}^{l}\frac{f'(t)}{\sqrt{R(t)}}\cot\frac{t-t_0}{a}dt,$$

last two terms on the right side of which tend to zero as $t_0 \to \pm l$. This is evident for the first one, and for the second one, it follows because the involved (principal) integral has logarithmic singularity while $\sqrt{R(t_0)}$ is an infinitesimal of order $\frac{1}{2}$.

Thus, we should have l such that

$$P_0 = \frac{4\mu}{(\kappa+1)\cos\frac{l}{a}}\int_{-l}^{l}\frac{f'(t)\tan\frac{t}{a}}{\sqrt{R(t)}}dt \tag{2.121}$$

is fulfilled. This equation may be explained as follows: if we wish press each stamp by periodic pressure force such that it is in contact with the boundary of the half-plane to an interval of length $2l$ in a period, then its magnitude P_0 should be given by (2.121). In the mean time, the pressure distribution beneath the stamp is

$$P(t_0) = \frac{4\mu\sqrt{R(t_0)}}{(\kappa+1)a\pi}\int_{-l}^{l}\frac{f'(t)}{\sqrt{R(t)}}\left(\cot\frac{t-t_0}{a} + \tan\frac{t}{a}\right)dt, \quad |t| < l. \tag{2.122}$$

As $a \to +\infty$ in (2.119) – (2.122), corresponding formulas for the case of a single stamp are obtained:

$$\Phi(z) = \frac{4\mu}{(\kappa+1)\pi\sqrt{l^2-z^2}}\int_{-l}^{l}\frac{f'(t)\sqrt{l^2-t^2}}{t-z}dt + \frac{P_0}{2\pi\sqrt{l^2-z^2}}, \quad z \in S^-, \tag{2.119'}$$

$$P(t_0) = \frac{4\mu}{(\kappa+1)\pi\sqrt{l^2-t_0^2}}\int_{-l}^{l}\frac{f'(t)\sqrt{l^2-t^2}}{t-t_0}dt + \frac{P_0}{2\pi\sqrt{l^2-t_0^2}}, \quad |t_0|< l, \tag{2.120$'$}$$

$$P_0 = \frac{4\mu}{\kappa+1}\int_{-l}^{l}\frac{tf'(t)}{\sqrt{l^2-t^2}}dt, \tag{2.121$'$}$$

$$P(t_0) = \frac{4\mu\sqrt{l^2-t_0^2}}{(\kappa+1)\pi}\int_{-l}^{l}\frac{f'(t)}{\sqrt{l^2-t^2}}\frac{dt}{t-t_0}, \quad |t_0|< l, \tag{2.122$'$}$$

which are identical to the results given in Muskhelishvili [1].

Example 2.8 *Periodic stamps with horizontal rectilinear base*
Here $f'(t)=0$. By (2.119) and (2.120),

$$\Phi(z) = \frac{P_0}{2a\pi\cos\frac{l}{a}\sqrt{\tan^2\frac{l}{a}-\tan^2\frac{z}{a}}} + \frac{\kappa-1}{\kappa+1}\frac{P_0}{2a\pi}$$

$$= \frac{P_0\cos\frac{l}{a}}{2a\pi\sqrt{\sin\frac{l+z}{a}\sin\frac{l-z}{a}}} + \frac{\kappa-1}{\kappa+1}\frac{P_0}{2a\pi},$$

$$P(t) = \frac{P_0}{a\pi\cos\frac{l}{a}\sqrt{\tan^2\frac{l}{a}-\tan^2\frac{t}{a}}} = \frac{P_0}{a\pi\sqrt{\sin\frac{l-t}{a}\sin\frac{l+t}{a}}},$$

where the radical involved is taken as the branch, when the plane is cut by γ, taking positive value as $z \to t \in \gamma_0$ from S^+.

Example 2.9 *Periodic stamps with inclined rectilinear base.*
Again assume the angle of inclination of the bases is ε (Fig.2.7). Then $f'(t)=\varepsilon$ in this case. By (2.119),

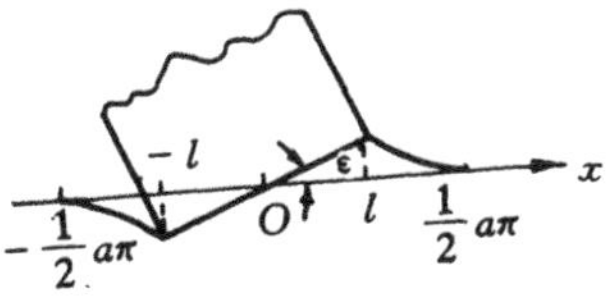

Fig.2.7

$$\Phi(z) = \frac{2\mu\varepsilon}{(\kappa+1)a\pi\sqrt{R(z)}}\int_{-l}^{l}\left(\cot\frac{t-z}{a}-\tan\frac{z}{a}\right)\sqrt{R(t)}\,dt$$

$$+ \frac{P_0}{2a\pi\cos\frac{l}{a}\sqrt{R(z)}} + \frac{\kappa-1}{\kappa+1}\frac{P_0}{2a\pi},$$

the first term on the right side of which is in fact

$$\frac{2\mu\varepsilon X(z)}{(\kappa+1)a\pi}\int_{\gamma_0}\left(\cot\frac{t-z}{a}-\tan\frac{z}{a}\right)\frac{dt}{X^+(t)}.$$

The integral appeared in it is exactly $I(z)$ expressed by (2.103) with 1 in place of κ. Thus, we have (here $A=0$)

$$\int_{\gamma_0}\left(\cot\frac{t-z}{a}-\tan\frac{z}{a}\right)\frac{dt}{X^+(t)}=a\pi i\left\{\frac{1}{X(z)}-\sec\frac{l}{a}\tan\frac{z}{a}\right\}$$

so that the mentioned term in $\Phi(z)$ becomes

$$\frac{2\mu\varepsilon}{\kappa+1}\left\{i-\frac{i\tan\frac{z}{a}X(z)}{\cos\frac{l}{a}}\right\}=\frac{2\mu\varepsilon}{\kappa+1}\left\{i-\frac{\tan\frac{z}{a}}{\cos\frac{l}{a}\sqrt{R(z)}}\right\}.$$

Substituting it into the expression of $\Phi(z)$, we obtain

$$\Phi(z)=\frac{2\mu\varepsilon}{\kappa+1}\left[i-\frac{\sin\frac{z}{a}}{\sqrt{\sin\frac{l+z}{a}\sin\frac{l-z}{a}}}\right]+\frac{P_0}{2a\pi}\left[\frac{\cos\frac{z}{a}}{\sqrt{\sin\frac{l+z}{a}\sin\frac{l-z}{a}}}+\frac{\kappa-1}{\kappa+1}\right].$$

In order to evaluate the pressure distribution right beneath the stamps, we may use (2.120). However, it would be more convenient by using the above formula directly.

In fact, for $-l<t<l$, we have

$$P(t)=\Phi^+(t)-\Phi^-(t)=-\frac{4\mu\varepsilon}{\kappa+1}\frac{\sin\frac{t}{a}}{\sqrt{\sin\frac{l+t}{a}\sin\frac{l-t}{a}}}+\frac{P_0}{a\pi}\frac{\cos\frac{t}{a}}{\sqrt{\sin\frac{l+t}{a}\sin\frac{l-t}{a}}}.$$

If regard the actual possibility in physics, i.e., $P(t)\geqslant 0$, it is sufficient for P_0 satisfying the condition

$$P_0\geqslant\frac{4\mu a\pi\varepsilon}{\kappa+1}\tan\frac{l}{a}.$$

Example 2.10 *Periodic stamps with circular base*

Assume the bases of periodic stamps are circular arcs of sufficiently large radius r, which may be taken approximately as

$$f(t) = \frac{a^2}{2r}\tan^2\frac{t}{a}, \quad -l \leqslant t \leqslant l$$

(The radius of curvature of this curve is r at $t=0$). Thus,

$$f'(t) = \frac{a}{r}\tan\frac{t}{a}\sec^2\frac{t}{a}.$$

Substituting it into (2.119), by symmetry, we have

$$\begin{aligned}\Phi(z) = & \frac{2\mu}{(\kappa+1)\pi r\sqrt{R(z)}}\int_{-l}^{l}\tan\frac{t}{a}\sec^2\frac{t}{a}\sqrt{R(t)}\cot\frac{t-z}{a}dt \\ & + \frac{P_0}{2a\pi}\left\{\frac{1}{\cos\frac{l}{a}\sqrt{R(z)}} + \frac{\kappa-1}{\kappa+1}\right\}.\end{aligned}$$

In order to evaluate the integral appeared on the right side, denoted by $J(z)$, making a transformation as in Example 2.7, we get

$$J(z) = a\int_{-L}^{L}u\sqrt{L^2-u^2}\,\frac{1+\zeta u}{u-\zeta}du = a(1+\zeta^2)\int_{-L}^{L}\frac{u\sqrt{L^2-u^2}}{u-\zeta}du$$

and then by the method used there, we have

$$J(z) = \frac{a\pi}{\cos^2\frac{z}{a}}\left\{i\tan\frac{z}{a}\sqrt{R(z)} + \frac{1}{2}\tan^2\frac{l}{a} - \tan^2\frac{z}{a}\right\}.$$

Substituting it into the above expression, after simplification, we obtain

$$\begin{aligned}\Phi(z) = & \frac{\mu a}{(\kappa+1)r\cos^2\frac{z}{a}}\left\{2i\tan\frac{z}{a} + \frac{\tan^2\frac{l}{a} - 2\tan^2\frac{z}{a}}{\sqrt{\tan^2\frac{l}{a} - \tan^2\frac{z}{a}}}\right\} \\ & + \frac{P_0}{2a\pi}\left\{\frac{1}{\cos\frac{l}{a}\sqrt{\tan^2\frac{l}{a} - \tan^2\frac{z}{a}}} + \frac{\kappa+1}{\kappa-1}\right\}.\end{aligned}$$

The pressure distribution beneath the stamp ($|t|<l$) is

$$\begin{aligned}P(t) & = \Phi^+(t) - \Phi^-(t) \\ & = \frac{2\mu a\left(\tan^2\frac{l}{a} - 2\tan^2\frac{t}{a}\right)}{(\kappa+1)r\cos^2\frac{l}{a}\sqrt{\tan^2\frac{l}{a} - \tan^2\frac{t}{a}}} + \frac{P_0}{a\pi\cos\frac{l}{a}\sqrt{\tan^2\frac{l}{a} - \tan^2\frac{t}{a}}}.\end{aligned}$$

For $P(t) \geqslant 0$,

$$P_0 \geqslant \frac{2\mu\pi a^2 \sin^2 \frac{l}{a}}{(\kappa + 1) r \cos^2 \frac{l}{a}}$$

must be fulfilled, which is certainly satisfied when r is large enough. If $P_0 > 0$ is given, by replacing the above inequality by equality, then, it may be verified that it has a unique solution l in $(0, \frac{1}{2} a\pi)$, which is the half width of the contact segment on the base in a period.

If let $a \to +\infty$ in the above three examples, we may obtain the corresponding results for the non-periodic case given in Muskhelishvili [1].

2. The case with friction

Now assume the friction coefficient $k \neq 0$ between the periodic stamps and the elastic half-plane, that means, beneath the stamps, between the shearing stress $T(t) = \tau_{xy}(t)$ and the normal pressure $P(t) = -\sigma_y(t)$, there exists the relation

$$T(t) = kP(t), \quad t \in \gamma_0. \tag{2.123}$$

Again assume $v^-(t) = f(t)$ on γ_0 with $f'(t) \in H$ and the external pressure force P_0 on γ_0 are given. The principal vector of the external stresses on γ_0 is known as $X + iY = T_0 - iP_0 =$
$(k - i)P_0$. On $\gamma_0' = L_0 - \gamma_0$, $T(t) = P(t) = 0$.

By (2.42) and (2.43), it is easy to obtain, when $t \in \gamma_0$,

$$(1 - ik)\Phi^+(t) + (1 + ik)\overline{\Phi}^+(t) = (1 - ik)\Phi^-(t) + (1 + ik)\overline{\Phi}^-(t), \tag{2.124}$$

$$\kappa\Phi^-(t) + \Phi^+(t) - \kappa\overline{\Phi}^+(t) - \overline{\Phi}^-(t) = 4i\mu f'(t), \tag{2.125}$$

and when $t \in \gamma_0'$,

$$\Phi^+(t) = \Phi^-(t).$$

From (2.124), we have

$$(1 - ik)\Phi(z) + (1 + ik)\overline{\Phi}(z) = 2\beta_2, \tag{2.126}$$

where β_2 is a constant. If $t_0 \in \gamma_0'$, we have

$$\overline{\Phi}(t) = \overline{\Phi(\bar{t})} = \overline{\Phi(t)}$$

on account of $\Phi^+(t) = \Phi^-(t) = \Phi(t)$. Thus, we know that β_2 is real.

Simplifying (2.125) by using (2.126) and eliminating $\overline{\Phi}(z)$, we arrive at the periodic Riemann boundary value problem

$$\frac{(\kappa+1)-ik(\kappa-1)}{1+ik}\Phi^+(t)+\frac{(\kappa+1)+ik(\kappa-1)}{1+ik}\Phi^-(t)$$
$$=4i\mu f'(t)+\frac{2(\kappa+1)\beta_2}{1+ik},$$

or, what is the same,

$$\Phi^+(t)=K\Phi^-(t)+f_0'(t)+\frac{2(\kappa+1)\beta_2}{(\kappa+1)-ik(\kappa-1)}, \tag{2.127}$$

where

$$K=-\frac{(\kappa+1)+ik(\kappa-1)}{(\kappa+1)-ik(\kappa-1)}, \tag{2.128}$$
$$f_0'(t)=\frac{4\mu i(1+ik)}{(\kappa+1)-ik(\kappa-1)}f'(t). \tag{2.129}$$

If put

$$\tan\pi\alpha=k\frac{\kappa-1}{\kappa+1}\quad(0<\alpha<\frac{1}{2}), \tag{2.130}$$

then, on account of

$$\kappa+1\pm ik(\kappa-1)=\frac{(\kappa+1)e^{\pm\pi\alpha i}}{\cos\pi\alpha}, \tag{2.131}$$

(2.128) and (2.129) may be rewritten respectively as

$$K=-e^{2\pi\alpha i}, \tag{2.128'}$$
$$f_0'(t)=\frac{4\mu i(1+ik)e^{\pi\alpha i}\cos\pi\alpha}{\kappa+1}f'(t), \tag{2.129'}$$

and the boundary value problem (2.127) as

$$\Phi^+(t)=-e^{2\pi\alpha i}\Phi^-(t)+f_0'(t)+2\beta_2e^{\pi\alpha i}\cos\pi\alpha. \tag{2.127'}$$

Now

$$\frac{\log K}{2\pi i}=\frac{1}{2}+\alpha. \tag{2.132}$$

The problem (2.127)′ should be solved in class h_0. Therefore, similar to the discussions on the particular example is § 2.2, Chapter I, it is easily seen that the index of the problem in class h_0 is 1 and its canonical function is

$$X(z)=(\tan\frac{z}{a}+\tan\frac{l}{a})^{-\frac{1}{2}-\alpha}(\tan\frac{z}{a}-\tan\frac{l}{a})^{-\frac{1}{2}+\alpha}, \tag{2.133}$$

holomorphic in the z-plane cut by γ, the branch of the function on the right side may be taken arbitrarily, for instance, such that

$$\lim_{z \to \pm \frac{1}{2} a\pi} \tan \frac{z}{a} X(z) = 1. \tag{2.134}$$

Analogous to the discussions in the previous paragraph, we obtain the general solution of (2.127)′ in h_0

$$\Phi(z) = \frac{2\mu(1 + ik)e^{\pi a i}\cos\pi a X(z)}{a\pi(\kappa + 1)} \int_{\gamma_0} \frac{f'(t)}{X^+(t)} \cot \frac{t - z}{a} dt$$
$$+ X(z)(1 + ik)i(\beta_0 \tan \frac{z}{a} + \beta_1) + \beta_2, \tag{2.135}$$

where β_0 and β_1 are two arbitrary constants.

We show that both β_0 and β_1 ought to be real so as condition (2.126) is satisfied.

First, we note that $X(\bar{z}) = \overline{X(z)}$, i.e., $\bar{X}(z) = X(z)$ since $X(z)$ takes real values when $z \in \gamma_0'$, and, when $t \in \gamma_0$,

$$\overline{X^+(t)} = \bar{X}^-(t) = X^-(t) = \frac{1}{K} X^+(t) = - e^{-2\pi a i} X^+(t).$$

There by

$$\frac{\Phi(z)}{1 + ik} = \frac{2\mu e^{\pi a i}\cos\pi a X(z)}{a\pi(\kappa + 1)} \int_{\gamma_0} \frac{f'(t)}{X^+(t)} \cot \frac{t - z}{a} dt$$
$$+ X(z)(\beta_0 \tan \frac{z}{a} + \beta_1)i + \frac{\beta_2}{1 + ik}$$

and

$$\frac{\bar{\Phi}(z)}{1 - ik} = \overline{\left[\frac{\Phi(\bar{z})}{1 + ik}\right]} = - \frac{2\mu e^{\pi a i}\cos\pi a X(z)}{a\pi(\kappa + 1)} \int_{\gamma_0} \frac{f'(t)}{X^+(t)} \cot \frac{t - z}{a} dt$$
$$- X(z)(\bar{\beta}_0 \tan \frac{z}{a} + \bar{\beta}_1)i + \frac{\beta_2}{1 - ik}.$$

To guarantee the validity of (2.126), i.e.,

$$\frac{\Phi(z)}{1 + ik} + \frac{\bar{\Phi}(z)}{1 - ik} = \frac{2\beta_2}{1 + k^2},$$

we see that both β_0 and β_1 are necessarily real.

As before, let us determine β_0, β_1 and β_2 by using the periodic condition of the displacements and the equilibrium condition of the stresses at $z = -\infty i$.

To this aim, consider Fig. 2.8 (in comparison with Fig. 2.5), where we have put

$$\zeta = \tan \frac{z}{a}, \quad L = \tan \frac{l}{a},$$
$$\theta = \arctan \frac{1}{L} = \frac{\pi}{2} - \frac{l}{a}, \quad 0 < \theta < \frac{\pi}{2}.$$

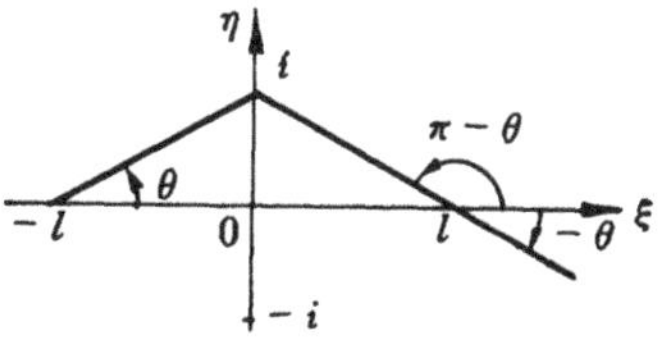

Fig. 2.8

Obviously

$$X(+\infty i) = X_*(i) = \cos\frac{l}{a}e^{(\frac{1}{2}-\alpha)\theta i}e^{(-\frac{1}{2}+\alpha)(\pi-\theta)i}$$
$$= -i\cos\frac{l}{a}e^{\alpha\pi i}e^{-2\alpha\theta i} = -i\cos\frac{l}{a}e^{\frac{2l\alpha}{a}i}. \qquad (2.136)$$

Similarly,

$$X(-\infty i) = i\cos\frac{l}{a}e^{-\frac{2l\alpha}{a}i}. \qquad (2.137)$$

Then, we have

$$\int_{\Lambda_\pm} X(z)dz = a\pi X(\pm\infty i) = \mp a\pi i\cos\frac{l}{a}e^{\pm\frac{2l\alpha}{a}i},$$
$$\int_{\Lambda_\pm} X(z)\tan\frac{z}{a}dz = \int_{\Lambda_\pm} X(z)\cot\frac{t-z}{a}dz$$
$$= \pm a\pi iX(\pm\infty i) = a\pi\cos\frac{l}{a}e^{\pm\frac{2l\alpha}{a}i},$$

where $\Lambda_\pm$ are the same as before. When $z\in S^-$, by (2.135), we have

$$\varphi(z+a\pi) - \varphi(z) = \int_{\Lambda_-}\Phi(z)dz$$
$$= \frac{2\mu(1+ik)\cos\pi\alpha\cos\frac{l}{a}e^{(\pi-\frac{2l}{a})\alpha i}}{\kappa+1}\int_{\gamma_0}\frac{f'(t)}{X^+(t)}dt$$
$$+ a\pi(1+ik)\cos\frac{l}{a}(i\beta_0-\beta_1)e^{-\frac{2l}{a}\alpha i} + a\pi\beta_2,$$
$$\varphi(\bar{z}+a\pi) - \varphi(\bar{z}) = \int_{\Lambda_+}\Phi(z)dz$$
$$= 2\mu(1+ik)\cos\pi\alpha\cos\frac{l}{a}e^{(\pi+\frac{2l}{a})\alpha i}\int_{\gamma_0}\frac{f'(t)}{X^+(t)}dt$$
$$+ a\pi(1+ik)\cos\frac{l}{a}(i\beta_0+\beta_1)e^{\frac{2l}{a}\alpha i} + a\pi\beta_2.$$

Thus, the condition of periodicity of the displacements is

$$i(e^{\frac{2la}{a}i} + \kappa e^{-\frac{2la}{a}i})\beta_0 + (e^{\frac{2la}{a}i} - \kappa e^{-\frac{2la}{a}i})\beta_1 + \frac{(\kappa+1)\beta_2}{(1+ik)\cos\frac{l}{a}}$$

$$= -\frac{2\mu\cos\pi\alpha}{(\kappa+1)a\pi}[e^{(\pi+\frac{2l}{a})\alpha i} + \kappa e^{(\pi-\frac{2l}{a})\alpha i}]\int_{\gamma_0}\frac{f'(t)}{X^+(t)}dt. \qquad (2.138)$$

Next, consider the condition of equilibrium of the stresses at $z = -\infty i$. It is obvious, by (2.136) and (2.137),

$$\Phi(-\infty i) = \frac{2\mu(1+ik)\cos\pi\alpha\cos\frac{l}{a}e^{(\pi-\frac{2l}{a})\alpha i}}{(\kappa+1)a\pi}\int_{\gamma_0}\frac{f'(t)}{X^+(t)}dt$$
$$+ (1+ik)\cos\frac{l}{a}(i\beta_0 - \beta_1)e^{-\frac{2la}{a}i} + \beta_2,$$

$$\Phi(+\infty i) = \frac{2\mu(1+ik)\cos\pi\alpha\cos\frac{l}{a}e^{(\pi+\frac{2l}{a})\alpha i}}{(\kappa+1)a\pi}\int_{\gamma_0}\frac{f'(t)}{X^+(t)}dt$$
$$+ (1+ik)\cos\frac{l}{a}(i\beta_0 + \beta_1)e^{\frac{2la}{a}i} + \beta_2,$$

and then we have, by noting that (2.112) is valid in this case,

$$\Phi(-\infty i) - \Phi(+\infty i) = \sigma_y(-\infty i) - i\tau_{xy}(-\infty i) = -\frac{(1+ik)P_0}{a\pi}.$$

Substituting the above expressions in it, we obtain

$$-\frac{P_0}{a\pi} = \frac{2\mu\cos\pi\alpha\cos\frac{l}{a}}{a\pi(\kappa+1)}[e^{(\pi-\frac{2l}{a})\alpha i} - e^{(\pi+\frac{2l}{a})\alpha i}]\int_{\gamma_0}\frac{f'(t)}{X^+(t)}dt$$
$$+ \cos\frac{l}{a}\{i\beta_0(e^{-\frac{2l}{a}\pi i} - e^{\frac{2l}{a}\pi i}) - \beta_1(e^{-\frac{2la}{a}i} + e^{\frac{2la}{a}i})\},$$

that is,

$$\beta_0\sin\frac{2la}{a} - \beta_1\cos\frac{2la}{a} = -\frac{2\mu\cos\pi\alpha\sin\frac{2la}{a}e^{\pi\alpha i}}{a\pi i(\kappa+1)}\int_{\gamma_0}\frac{f'(t)}{X^+(t)}dt - \frac{P_0}{2a\pi\cos\frac{l}{a}}. \qquad (2.139)$$

Put

$$Q(z) = (\tan\frac{l}{a} + \tan\frac{z}{a})^{\frac{1}{2}+\alpha}(\tan\frac{l}{a} - \tan\frac{z}{a})^{\frac{1}{2}-\alpha}, \qquad (2.140)$$

and choose its branch in the plane cut by γ such that it takes positive value Q^+

$(t) = Q(t)$ as $z \to t \in \gamma_0$ from S^+. And so, on account of

$$\arg X^+(t) = \left(-\frac{1}{2} + \alpha\right)\pi, \; t \in \gamma_0,$$

we have

$$X(z) = -\frac{ie^{\alpha\pi i}}{Q(z)},$$
$$X^+(t) = -\frac{ie^{\alpha\pi i}}{Q(t)}, \; t \in \gamma_0,$$

in which $Q(t) > 0$. Thus, (2.135) may be rewritten as

$$\Phi(z) = \frac{2\mu(1+ik)e^{\pi\alpha i}\cos\pi\alpha}{a\pi(\kappa+1)Q(z)}\int_{\gamma_0} f'(t)Q(t)\cot\frac{t-z}{a}dt + \frac{(1+ik)e^{\pi\alpha i}}{Q(z)}\left(\beta_0\tan\frac{z}{a} + \beta_1\right) + \beta_2. \qquad (2.135)'$$

Thus, β_0, β_1 and β_2 may be determined by (2.138) and (2.139) or by the following two equations:

$$i\left(e^{\frac{2l\alpha}{a}i} + \kappa e^{-\frac{2l\alpha}{a}i}\right)\beta_0 + \left(e^{\frac{2l\alpha}{a}i} - \kappa e^{-\frac{2l\alpha}{a}i}\right)\beta_1 + \frac{\kappa+1}{(1+ik)\cos\frac{l}{a}}\beta_2 = \frac{2\mu\cos\pi\alpha}{(\kappa+1)a\pi i}\left(e^{\frac{2l}{a}\alpha i} + \kappa e^{-\frac{2l\alpha}{a}i}\right)\int_{\gamma_0} f'(t)Q(t)dt, \qquad (2.138)'$$

$$\beta_0\sin\frac{2l\alpha}{a} - \beta_1\cos\frac{2l\alpha}{a} = -\frac{2\mu\cos\pi\alpha\sin\frac{2l\alpha}{a}}{(\kappa+1)a\pi}\int_{\gamma_0} f'(t)Q(t)dt - \frac{P_0}{2a\pi\cos\frac{l}{a}}. \qquad (2.139)'$$

Seperating the real and imaginary parts in (2.138)′, we get

$$\beta_0\cos\frac{2l\alpha}{a} + \beta_1\sin\frac{2l\alpha}{a} - \frac{k}{k^2+1}\frac{\beta_2}{\cos\frac{l}{a}} = -\frac{2\mu\cos\pi\alpha\cos\frac{2l\alpha}{a}}{(\kappa+1)a\pi}\int_{\gamma_0} f'(t)Q(t)dt, \qquad (2.141)$$

$$\beta_0\sin\frac{2l\alpha}{a} - \beta_1\cos\frac{2l\alpha}{a} + \frac{\kappa+1}{\kappa-1}\frac{\beta_2}{(k^2+1)\cos\frac{l}{a}} = -\frac{2\mu\cos\pi\alpha\sin\frac{2l\alpha}{a}}{(\kappa+1)a\pi}\int_{\gamma_0} f'(t)Q(t)dt. \qquad (2.142)$$

Comparing (2.142) with (2.139)′, we then obtain

$$\beta_2 = \frac{\kappa - 1}{\kappa + 1}\frac{(k^2 + 1)P_0}{2a\pi}. \tag{2.143}$$

By substituting it into (2.141) and combining with (2.139)′, β_0 and β_1 may be found:

$$\beta_0 = -\frac{2\mu\cos\pi\alpha}{(\kappa + 1)a\pi}\int_{\gamma_0} f'(t)Q(t)dt + \frac{P_0\sin(1 - \lambda)\pi\alpha}{2a\pi\cos\pi\alpha\cos\frac{l}{a}}, \tag{2.144}$$

$$\beta_1 = \frac{P_0\cos(1 - \lambda)\pi\alpha}{2a\pi\cos\pi\alpha\cos\frac{l}{a}}, \tag{2.145}$$

where we have set

$$\lambda = \frac{2l}{a\pi} \quad (0 < \lambda < 1). \tag{2.146}$$

By substituting (2.143) – (2.145) into (2.135)′, the expression of $\Phi(z)$ may be obtained at length.

Thus, we may calculate the pressure distribution beneath the stamps. Since we have

$$P(t_0) + iT(t_0) = (1 + ik)P(t_0) = \Phi^+(t_0) - \Phi^-(t_0),$$

using the generalized Plemelj formula and noting that

$$Q^-(t) = -Q(t)e^{2\alpha\pi i}, \quad t \in \gamma_0, \tag{2.147}$$

we get

$$P(t_0) = -\frac{2\mu\sin 2\pi\alpha}{\kappa + 1}f'(t_0) + \frac{4\mu\cos^2\pi\alpha}{a\pi(\kappa + 1)Q(t_0)}\int_{-l}^{l} f'(t)Q(t)\cot\frac{t - t_0}{a}dt$$
$$+ \frac{2\cos\pi\alpha}{Q(t_0)}\left(\beta_0\tan\frac{t_0}{a} + \beta_1\right), \quad t_0 \in \gamma_0, \tag{2.148}$$

where β_0 and β_1 are given by (2.144) and (2.145) respectively.

When $k = 0$ (consequently $\alpha = 0$), we obtain the results given in the previous paragraph for the case without friction. On the other hand, if let $a \to +\infty$, Muskhelishvili's results in [1] for the non-periodic case may be obtained.

Example 2.11 *Periodic stamps with horizontal rectilinear base*
In this case, $f'(t) = 0$.
Then, by (2.135) and (2.143) – (2.145), we get

$$\Phi(z) = \frac{P_0(1 + ik)e^{\pi\alpha i}\cos\left[\frac{z}{a} - (1 - \lambda)\pi\alpha\right]}{2a\pi\cos\pi\alpha\sin^{\frac{1}{2}+\alpha}\frac{l + z}{a}\sin^{\frac{1}{2}-\alpha}\frac{l - z}{a}} + \frac{\kappa - 1}{\kappa + 1}\frac{(k^2 + 1)P_0}{2a\pi}, \tag{2.149}$$

where, in the z-plane cut by γ, $\sin^{\frac{1}{2}+\alpha}\frac{l+z}{a}\sin^{\frac{1}{2}-\alpha}\frac{l-z}{a}$ has been chosen to be the branch which takes positive value as $z \to t \in \gamma_0$ from S^+, in which λ is still given by (2.146).

By (2.148), the pressure distribution on the boundary beneath the stamps is given by

$$P(t) = \frac{P_0 \cos[\frac{t}{a} - (1-\lambda)\pi\alpha]}{a\pi \sin^{\frac{1}{2}+\alpha}\frac{l+t}{a}\sin^{\frac{1}{2}-\alpha}\frac{l-t}{a}}, \quad t \in \gamma_0. \tag{2.150}$$

If $k=0$ (then $\alpha=0$), it is the result given in Example 2.6 for the case without friction.

Example 2.12 *Periodic stamps with inclined rectilinear base*

Assume again the angle of inclination is ε, then $f'(t)=\varepsilon$.

Then, by the related equalities, we have

$$\Phi(z) = \frac{2\mu\varepsilon(1+ik)e^{\pi\alpha i}\cos\pi\alpha}{a\pi(\kappa+1)Q(z)}\int_{\gamma_0} Q(t)(\cot\frac{t-z}{a} - \tan\frac{z}{a})dt + \Phi_1(z), \tag{2.151}$$

where $\Phi_1(z)$ is exactly the function $\Phi(z)$ in Example 2.11.

Let us evaluate the integral involved in the above equation:

$$I(z) = \int_{\gamma_0} Q(t)(\cot\frac{t-z}{a} - \tan\frac{z}{a})dt.$$

As before, let $u=\tan\frac{t}{a}$, $\zeta=\tan\frac{z}{a}$. $L=\tan\frac{l}{a}$. Then, it is easily verified that

$$I(z) = a\sec^2\frac{z}{a}J(\zeta),$$

where

$$J(\zeta) = \int_{-L}^{L}\frac{(L+u)^{\frac{1}{2}+\alpha}(L-u)^{\frac{1}{2}-\alpha}}{(1+u^2)(u-\zeta)}du.$$

Set

$$\Omega(\zeta) = \int_{\Lambda}\frac{(L+w)^{\frac{1}{2}+\alpha}(L-w)^{\frac{1}{2}-\alpha}}{(1+w^2)(w-\zeta)}dw,$$

where Λ is a closed contour surrounding the line-segment $[-L, L]$ in the w-plane but leaving ζ in its exterior and the path of integration is taken clockwisely. On account of (2.147), we assure that

$$\Omega(\zeta) = (1 + e^{2\alpha\pi i})J(\zeta),$$

so that it is sufficient to evaluate $\Omega(\zeta)$.

At $w=\infty$, the integrand in the above expresson of $\Omega(\zeta)$ has a double zero-point, and at $w=i$ (see Fig. 2.8),

$$\begin{aligned}(L + w)^{\frac{1}{2}+\alpha}(L - w)^{\frac{1}{2}-\alpha}\big|_{w=i} &= |L + i|^{\frac{1}{2}+\alpha}|L - i|^{\frac{1}{2}-\alpha}e^{(\frac{1}{2}+\alpha)\theta i}e^{(-\frac{1}{2}+\alpha)\theta i} \\ &= \sec\frac{l}{a}e^{2\alpha\theta i} = \sec\frac{l}{a}e^{2\alpha(\frac{\pi}{2}-\frac{l}{a})i} \\ &= \sec\frac{l}{a}e^{(1-\lambda)\alpha\pi i},\end{aligned}$$

and at $w=-i$,

$$\begin{aligned}(L + w)^{\frac{1}{2}+\alpha}(L - w)^{\frac{1}{2}-\alpha}\big|_{w=-i} &= \sec\frac{l}{a}e^{-(\frac{1}{2}+\alpha)\theta i}e^{(\frac{1}{2}-\alpha)(-2\pi+\theta)i} \\ &= -\sec\frac{l}{a}e^{2\pi\alpha i}e^{-2\alpha\theta i} = -\sec\frac{l}{a}e^{(1+\lambda)\alpha\pi i}.\end{aligned}$$

Thus, by the well-know residue theorem, we get

$$\frac{1}{2\pi i}\Omega(\zeta) = \frac{(L + \zeta)^{\frac{1}{2}+\alpha}(L - \zeta)^{\frac{1}{2}-\alpha}}{1 + \zeta^2} - \frac{e^{(1-\lambda)\alpha\pi i}}{2i(\zeta - i)\cos\frac{l}{a}} - \frac{e^{(1+\lambda)\alpha\pi i}}{2i(\zeta + i)\cos\frac{l}{a}}.$$

Therefore, we obtain

$$\Omega(\zeta) = \frac{2\pi i(L + \zeta)^{\frac{1}{2}+\alpha}(L - \zeta)^{\frac{1}{2}-\alpha}}{1 + \zeta^2} - \frac{2\pi e^{\alpha\pi i}(\zeta\cos\lambda\alpha\pi + \sin\lambda\alpha\pi)}{(1 + \zeta^2)\cos\frac{l}{a}}.$$

Since

$$I(z) = \frac{a}{\cos^2\frac{z}{a}}\frac{1}{1 + e^{2\alpha\pi i}}\Omega(\zeta) = \frac{ae^{-\alpha\pi i}}{2\cos^2\frac{z}{a}\cos\alpha\pi}\Omega(\zeta),$$

finally we have

$$I(z) = \frac{\alpha\pi i Q(z)}{\cos\alpha\pi e^{\alpha\pi i}} - \frac{a\pi\sin(\frac{z}{a} + \lambda\alpha\pi)}{\cos\alpha\pi\cos\frac{l}{a}\cos\frac{z}{a}}.$$

Substituting it into (2.151), we obtain

$$\Phi(z) = \frac{2\mu\varepsilon i(1 + ik)}{\kappa + 1} - \frac{2\mu\varepsilon(1 + ik)e^{\alpha\pi i}\sin(\frac{z}{a} + \lambda\alpha\pi)}{(\kappa + 1)\sin^{\frac{1}{2}+\alpha}\frac{l + z}{a}\sin^{\frac{1}{2}-\alpha}\frac{l - z}{a}} + \Phi_1(z), \tag{2.152}$$

where $\Phi_1(z)$ is given by the right-hand member of (2.149).

Let us now calculate the stress distribution beneath the stamps.

Without using (2.149), we may calculate it directly and rapidly. As

$$P(t) = \frac{\Phi^+(t) - \Phi^-(t)}{1 + ik},$$

by noting (2.147), the expression of $P(t)$ is readily obtained:

$$P(t) = -\frac{4\mu\varepsilon\cos\alpha\pi\sin(\frac{t}{a} + \lambda\alpha\pi)}{(\kappa + 1)\sin^{\frac{1}{2}+\alpha}\frac{l + t}{a}\sin^{\frac{1}{2}-\alpha}\frac{l - t}{a}} + P_1(t),$$

where $P_1(t)$ is given by the right-hand member of (2.150).

Thus, we find at length

$$P(t) = \frac{P_0\cos[\frac{t}{a} - (1 - \lambda)\alpha\pi]}{a\pi\sin^{\frac{1}{2}+\alpha}\frac{l + t}{a}\sin^{\frac{1}{2}-\alpha}\frac{l - t}{a}} - \frac{4\mu\varepsilon\cos\alpha\pi\sin(\frac{t}{a} - \lambda\alpha\pi)}{(\kappa + 1)\sin^{\frac{1}{2}+\alpha}\frac{l + t}{a}\sin^{\frac{1}{2}-\alpha}\frac{l - t}{a}}, \quad t \in \gamma_0. \tag{2.153}$$

Obviously, if $\varepsilon > 0$ is sufficiently small, for example, if

$$\varepsilon \leqslant \frac{P_0(\kappa + 1)\cos[(\alpha + \frac{1}{2})\lambda\pi - \alpha\pi]}{4\mu\cos\alpha\pi\sin(\alpha + \frac{1}{2})\lambda\pi},$$

then

$$P(t) \geqslant 0$$

is guaranteed, that means, it is actually possible in physics.

If put $k = 0$ in (2.152) and (2.153), then the results in Example 2.9 for the case without friction are reached. If let $a \to +\infty$, then the results of Muskhelishivili [1] for the non-periodic case are achieved.

Remark 1 By applying the method used by Muskhelishvili and that similar to the present section, the more general periodic contact problems (e.g., the contact line not necessarily rectilinear) in isotropic plane elasticity may be solved, which would not be discussed here.

Remark 2 It is assumed only a single stamp appeared in one period in our above discussions. However, it may be extended to the case where there are several stamps in a period, without any difficulty in principle.

Chapter Ⅲ

Periodic Problems for Anisotropic Medium

In the present chapter, theory of periodic problems in anisotropic plane elasticity will be discussed.

Under the assumptions that stresses and displacements are periodic and the stresses at infinity are bounded, the periodicity of the stress functions will be studied. Then, by using the formula of integrals with Hilbert kernel, the fundamental problems of anisotropic elastic half-plane will be obtained. At last, the periodic contact problems in anisotropic plane elasticity will be investigated, which could be reduced to periodic Riemann-Hilbert boundary value problems by introducing two functions defined in terms of certain integrals with Hilbert kernel.

§ 1. The Stress Functions

1. Basic assumptions

All the discussions below are under the following assumptions: the medium of the $a\pi$-periodic region is anisotropic, the stresses and displacements are periodic and the stresses at infinity are bounded, and the involved boundary conditions are periodic. Thus, we need only restrict our discussions in a period part of the elastic body. Moreover, we assume the elastic body occupies the lower half-plane in the z-plane ($z = x + iy$) and so there is only one point at infinity $z = -\infty i$.

The principal vector $X(-\infty i) + iY(-\infty i)$ of the external stresses at $z = -\infty i$ is understood by the limit of the principal vector of the stresses along a line-segment from z to $z + a\pi$ in S^- as $z \to -\infty i$, i.e.,

$$X(-\infty i) = a\pi\tau_{xy}(-\infty i), \quad Y(-\infty i) = a\pi\sigma_y(-\infty i).$$

If the principal vector of the external stresses on the boundary in a period is $X + iY$, then, by condition of equilibrium,

$$X + iY = X(-\infty i) + iY(-\infty i),$$

that is

$$\sigma_y(-\infty i) = \frac{Y}{a\pi}, \quad \tau_{xy}(-\infty i) = \frac{X}{a\pi}. \tag{3.1}$$

2. Periodicity of the stress functions for anisotropic medium

For *anisotropic elastic body*, the stress components σ_x, σ_y, τ_{xy} and displacement components u, v may be expressed by means of $\varphi(z_1)$ and $\psi(z_2)$ or their derivatives $\Phi(z_1) = \varphi'(z_1)$ and $\Psi(z_2) = \psi'(z_2)$ (*stress functions*):

$$\sigma_x = \mu_1^2\Phi(z_1) + \bar{\mu}_1^2\,\overline{\Phi(z_1)} + \mu_2^2\Psi(z_2) + \bar{\mu}_2^2\,\overline{\Psi(z_2)}, \tag{3.2}$$

$$\sigma_y = \Phi(z_1) + \overline{\Phi(z_1)} + \Psi(z_2) + \overline{\Psi(z_2)}, \tag{3.3}$$

$$\tau_{xy} = -\left[\mu_1\Phi(z_1) + \bar{\mu}_1\,\overline{\Phi(z_1)} + \mu_2\Psi(z_2) + \bar{\mu}_2\,\overline{\Psi(z_2)}\right], \tag{3.4}$$

$$u = p_1\varphi(z_1) + \bar{p}_1\,\overline{\varphi(z_1)} + p_2\psi(z_2) + \bar{p}_2\,\overline{\psi(z_2)}, \tag{3.5}$$

$$v = q_1\varphi(z_1) + \bar{q}_1\,\overline{\varphi(z_1)} + q_2\psi(z_2) + \bar{q}_2\,\overline{\psi(z_2)}, \tag{3.6}$$

where $\varphi(z_1)$ and $\psi(z_2)$ are functions holomorphic in z_1 and z_2 respectively, in which

$$z_1 = x + \mu_1 y, \quad z_2 = x + \mu_2 y,$$

and

$$\left.\begin{aligned} p_1 &= \beta_{11}\mu_1^2 + \beta_{12} - \beta_{16}\mu_1, \\ p_2 &= \beta_{11}\mu_2^2 + \beta_{12} - \beta_{16}\mu_2, \\ q_1 &= \frac{\beta_{12}\mu_1^2 + \beta_{22} - \beta_{26}\mu_1}{\mu_1}, \\ q_2 &= \frac{\beta_{12}\mu_2^2 + \beta_{22} - \beta_{26}\mu_2}{\mu_2}, \end{aligned}\right\} \tag{A}$$

and μ_1, $\bar{\mu}_1$, μ_2, $\bar{\mu}_2$ are the roots of the equation

$$\beta_{11}s^4 - 2\beta_{16}s^3 + (2\beta_{12} + \beta_{66})s^2 - 2\beta_{26}s + \beta_{22} = 0,$$

while

$$\begin{matrix} \beta_{11} & \beta_{12} & \beta_{16} \\ \beta_{12} & \beta_{22} & \beta_{26} \\ \beta_{16} & \beta_{26} & \beta_{66} \end{matrix}$$

are the elastic coefficients of the anisotropic elastic body (see Savin [1]).

Lemma 3.1 *Under the basic assumptions, the stress functions* $\Phi(z_1)$ *and* $\Psi(z_2)$ *are* $a\pi$*-periodic functions.*

Proof Taking partial derivatives with respect to x in (3.5) and (3.6), we get

$$\frac{\partial u}{\partial x} = p_1\Phi(z_1) + \bar{p}_1\,\overline{\Phi(z_1)} + p_2\Psi(z_2) + \bar{p}_2\,\overline{\Psi(z_2)}, \tag{3.7}$$

$$\frac{\partial v}{\partial x} = q_1\Phi(z_1) + \bar{q}_1\overline{\Phi(z_1)} + q_2\Psi(z_2) + \bar{q}_2\overline{\Psi(z_2)}. \tag{3.8}$$

By the periodicity (even quasi-periodicity) of the displacements, we know that $\frac{\partial u}{\partial x}$ and $\frac{\partial v}{\partial x}$ are $a\pi$- periodic.

Since the coefficient determinant of the system of linear equations (3.2), (3.3), (3.7) and (3.8)

$$\begin{vmatrix} \mu_1^2 & \bar{\mu}_1^2 & \mu_2^2 & \bar{\mu}_2^2 \\ 1 & 1 & 1 & 1 \\ p_1 & \bar{p}_1 & p_2 & \bar{p}_2 \\ q_1 & \bar{q}_1 & q_2 & \bar{q}_2 \end{vmatrix} \neq 0,$$

we may obtain its unique solution including $\Phi(z_1)$, $\Psi(z_2)$ in terms of $\frac{\partial u}{\partial x}$, $\frac{\partial v}{\partial x}$, σ_x and σ_y, which are periodic.

§ 2. Periodic Fundamental Problems of Anisotropic Half-plane

In the present section, periodic fundamental problems of anisotropic elastic half-plane will be discussed and directly solved by using integrals with Hilbert kernel.

1. The first fundamental problem

Assume the anisotropic body occupies the lower half-plane S^- of the z-plane. Denote $z = t$ (real) on the x-axis. Given the external stresses on the x-axis:

$$\sigma_y(t) = -P(t), \quad \tau_{xy}(t) = T(t), \tag{3.9}$$

which are arc-wisely Hölder continuous and periodic. Under the basic assumptions, find the stress distribution (and displacements), called the *periodic first fundamental problem*.

Theorem 3.1 *Under the above assumptions, the first fundamental problem of the anisotropic half-plane is uniquely solvable.*

We would give its method of solution simultaneously in the proof of the theorem.

Proof First of all, using (3.3) and (3.4), take the linear combinations of σ_y and τ_{xy}: $\mu_2\sigma_y + \tau_{xy}$ and $\mu_1\sigma_y + \tau_{xy}$. Then, let z_1 and z_2 tend to t on the x-axis from S^-. On account of (3.9), we get

$$-\mu_2 P(t) + T(t) = (\mu_2 - \mu_1)\Phi(t) + (\mu_2 - \bar{\mu}_1)\overline{\Phi(t)} + (\mu_2 - \bar{\mu}_2)\overline{\Psi(t)},$$

$$-\mu_1P(t)+T(t)=(\mu_1-\bar{\mu}_1)\overline{\Phi(t)}+(\mu_1-\mu_2)\Psi(t)+(\mu_1-\bar{\mu}_2)\overline{\Psi(t)}.$$

Multiplying both sides of these two equations by $\frac{1}{2a\pi i}\cot\frac{t-z_1}{a}dt$ and $\frac{1}{2a\pi i}\cot\frac{t-z_2}{a}dt$ respectively and integrating along L_0: $-\frac{a\pi}{2}\leqslant t\leqslant\frac{a\pi}{2}$, and applying the formulas (1.59) and (1.60) for integrals with Hilbert kernel, we obtain

$$\Phi(z_1)=\frac{-1}{(\mu_1-\mu_2)2a\pi i}\int_{L_0}[\mu_2P(t)-T(t)]\cot\frac{t-z_1}{a}dt+\gamma_1, \tag{3.10}$$

$$\Psi(z_2)=\frac{1}{(\mu_1-\mu_2)2a\pi i}\int_{L_0}[\mu_1P(t)-T(t)]\cot\frac{t-z_2}{a}dt+\gamma_2, \tag{3.11}$$

where

$$\gamma_1=-\frac{1}{2}\frac{1}{\mu_1-\mu_2}[(\mu_2-\mu_1)\Phi(-\infty i) + (\mu_2-\bar{\mu}_1)\overline{\Phi(-\infty i)}+(\mu_2-\bar{\mu}_2)\overline{\Psi(-\infty i)}], \tag{3.12}$$

$$\gamma_2=\frac{1}{2}\frac{1}{\mu_1-\mu_2}[(\mu_1-\mu_2)\Psi(-\infty i) - (\mu_1-\bar{\mu}_1)\overline{\Phi(-\infty i)}-(\mu_1-\bar{\mu}_2)\overline{\Psi(-\infty i)}]. \tag{3.13}$$

In order to determine γ_1 and γ_2, by noting (3.12) and (3.13), it is easily seen that

$$\mathrm{Re}\{\gamma_1+\gamma_2\}=0, \tag{3.14}$$

$$\mathrm{Re}\{\mu_1\gamma_1+\mu_2\gamma_2\}=0.\text{[①]} \tag{3.15}$$

Then, by using the periodic condition of the displacements, we may arrive at another two equations in γ_1 and γ_2.

To this aim, integrating (3.10) and (3.11), we get, neglecting a rigid translation,

$$\varphi(z_1)=\frac{1}{(\mu_1-\mu_2)2\pi i}\int_{L_0}(\mu_2P-T)\log\sin\frac{t-z_1}{a}dt+\gamma_1z_1,$$

$$\psi(z_2)=\frac{-1}{(\mu_1-\mu_2)2\pi i}\int_{L_0}(\mu_1P-T)\log\sin\frac{t-z_2}{a}dt+\gamma_2z_2.$$

Then, for $z\in S^-$, i.e., $z_1, z_2\in S^-$,

$$\varphi(z_1+a\pi)=\frac{1}{(\mu_1-\mu_2)2\pi i}\int_{L_0}(\mu_2P-T)[\log\sin\frac{t-z_1}{a}$$

① (3.14) and (3.15) may be also verified by the condition of equilibrium at $z=-\infty i$.

$$+ i\pi]dt + \gamma_1(z_1 + a\pi),$$

$$\psi(z_2 + a\pi) = \frac{-1}{(\mu_1 - \mu_2)2\pi i}\int_{L_0}(\mu_1 P - T)[\log\sin\frac{t - z_2}{a} + i\pi]dt + \gamma_2(z_2 + a\pi),$$

and consequently

$$\varphi(z_1 + a\pi) - \varphi(z_1) = \frac{1}{2(\mu_1 - \mu_2)}\int_{L_0}(\mu_2 P - T)dt + \gamma_1 a\pi,$$

$$\psi(z_2 + a\pi) - \psi(z_2) = \frac{-1}{2(\mu_1 - \mu_2)}\int_{L_0}(\mu_1 P - T)dt + \gamma_2 a\pi.$$

Therefore, by using (3.5) and (3.6), the increments of u and v on L_0 may be evaluated:

$$u\,|_{L_0} = \mathrm{Re}\left\{\frac{p_1\mu_2 - p_2\mu_1}{\mu_1 - \mu_2}\int_{L_0}Pdt - \frac{p_1 - p_2}{\mu_1 - \mu_2}\int_{L_0}Tdt + 2a\pi(p_1\gamma_1 + p_2\gamma_2)\right\},$$

$$v\,|_{L_0} = \mathrm{Re}\left\{\frac{q_1\mu_2 - q_2\mu_1}{\mu_1 - \mu_2}\int_{L_0}Pdt - \frac{q_1 - q_2}{\mu_1 - \mu_2}\int_{L_0}Tdt + 2a\pi(q_1\gamma_1 + q_2\gamma_2)\right\}.$$

Thus, by the periodicity of the displacements: $u\,|_{L_0} = v\,|_{L_0} = 0$, the following equations are obtained:

$$B_1 P^* - B_2 T^* + 2\mathrm{Re}\{p_1\gamma_1 + p_2\gamma_2\} = 0, \quad (3.16)$$

$$B_3 P^* - B_4 T^* + 2\mathrm{Re}\{q_1\gamma_1 + q_2\gamma_2\} = 0, \quad (3.17)$$

where we have put

$$P^* = \frac{1}{a\pi}\int_{L_0}Pdt, \quad T^* = \frac{1}{a\pi}\int_{L_0}Tdt,$$

$$\left.\begin{aligned}
B_1 &= \mathrm{Re}\left\{\frac{p_1\mu_2 - p_2\mu_1}{\mu_1 - \mu_2}\right\} = \mathrm{Re}\{\beta_{11}\mu_1\mu_2 - \beta_{12}\},\\
B_2 &= \mathrm{Re}\left\{\frac{p_1 - p_2}{\mu_1 - \mu_2}\right\} = \mathrm{Re}\{\beta_{11}(\mu_1 + \mu_2) - \beta_{26}\},\\
B_3 &= \mathrm{Re}\left\{\frac{q_1\mu_2 - q_2\mu_1}{\mu_1 - \mu_2}\right\} = \mathrm{Re}\{-\beta_{22}(\frac{1}{\mu_1} + \frac{1}{\mu_2}) + \beta_{26}\},\\
B_4 &= \mathrm{Re}\left\{\frac{q_1 - q_2}{\mu_1 - \mu_2}\right\} = \mathrm{Re}\{\beta_{12} - \beta_{22}\frac{1}{\mu_1\mu_2}\}.
\end{aligned}\right\} \quad (*)$$

Hence, γ_1 and γ_2 may be determined by (3.14) – (3.17), and then $\Phi(z_1)$, $\Psi(z_2)$ may be obtained. Thereby, the stress distribution of the elastic body is determined.

Let us clarify some special case used to occur in engineering.

Consider the case of *orthotropic elastic body*, with one of the orthotropic axes parallel to the direction of the period, and the other, perpendicular to it. In this case, $\mu_1 = i\nu_1$, $\mu_2 = i\nu_2$, $\beta_{16} = \beta_{26} = 0$. Thus, by (A) and (*), we have

$$p_1 = -\nu_1^2\beta_{11} + \beta_{12} = G_1,\ p_2 = -\nu_2^2\beta_{11} + \beta_{12} = G_2,$$
$$q_1 = (\beta_{12}\nu_1 - \frac{\beta_{22}}{\nu_1})i = H_1 i,\ q_2 = (\beta_{12}\nu_2 - \frac{\beta_{22}}{\nu_2})i = H_2 i,$$
$$B_1 = -(\beta_{11}\nu_1\nu_2 + \beta_{12}),\ B_2 = B_3 = 0,\ B_4 = \beta_{12} + \frac{\beta_{22}}{\nu_1\nu_2},$$

where ν_1, ν_2, G_1, G_2, H_1, H_2 are all real numbers. At the same time,

$$\Phi(z_1) = \frac{1}{(\nu_1 - \nu_2)2a\pi}\int_{L_0}(i\nu_2 P - T)\cot\frac{t - z_1}{a}dt + \gamma_1,$$
$$\Psi(z_2) = \frac{-1}{(\nu_1 - \nu_2)2a\pi}\int_{L_0}(i\nu_1 P - T)\cot\frac{t - z_2}{a}dt + \gamma_2.$$

The equations (3.14) – (3.17) for determining γ_1 and γ_2 become

$$\gamma_1^* + \gamma_2^* = 0,\ \nu_1\gamma_1' + \nu_2\gamma_2' = 0,$$
$$G_1\gamma_1^* + G_2\gamma_2^* = -\frac{1}{2}B_1P^*,\ H_1\gamma_1' + H_2\gamma_2' = -\frac{1}{2}B_4P^*,$$

where γ_j^* and γ_j' are the real and the imaginary parts of γ_j $(j = 1, 2)$ respectively.

After $\gamma_1 = \gamma_1^* + i\gamma_1'$ and $\gamma_2 = \gamma_2^* + i\gamma_2'$ are determined, we obtain finally

$$\Phi(z_1) = \frac{1}{2a\pi(\nu_1 - \nu_2)}\int_{L_0}(i\nu_2 P - T)\cot\frac{t - z_1}{a}dt - \frac{B_1P^*}{2(G_1 - G_2)} - \frac{iB_4T^*\nu_2}{2(H_1\nu_2 - H_2\nu_1)}, \tag{3.18}$$
$$\Psi(z_2) = \frac{-1}{2a\pi(\nu_1 - \nu_2)}\int_{L_0}(i\nu_1 P - T)\cot\frac{t - z_2}{a}dt + \frac{B_1P^*}{2(G_1 - G_2)} + \frac{iB_4T^*\nu_1}{2(H_1\nu_2 - H_2\nu_1)}. \tag{3.19}$$

Here, P^* and T^* are also the external pressure force and shearing force at $z = -\infty i$ respectively. In fact, substituting (3.18) and (3.19) into (3.3) and (3.4), and letting $z \to -\infty i$, we get

$$\sigma_y(-\infty i) = -P^*, \tag{3.20}$$
$$\tau_{xy}(-\infty i) = T^*, \tag{3.21}$$

which are also obvious by the condition of equilibrium. Similarly, we have

$$\sigma_x(-\infty i) = \frac{\beta_{12}}{\beta_{11}}P^*. \tag{3.22}$$

We see that $\sigma_x(-\infty i) < 0$ when $P^* > 0$ since $\beta_{11} > 0$ and $\beta_{12} < 0$. Therefore, we assure that, in our case, if the resultant of the normal loads on the segment in a period of the boundary is pressure: $\int_{L_0} P dt = a\pi P^* > 0$, then $\sigma_x(-\infty i)$ is also a pressure, determined by (3.22).

If the elastic body is isotropic, then (2.73) may be obtained by (3.22).

Example 3.1 Assume that, on a sub-segment of the boundary of the orthotropic elastic half-plane in a period, a uniform pressure is applied, i.e., in a period,

$$P(t) = \begin{cases} P, & |t| \leqslant l, \\ 0, & l < |t| \leqslant \dfrac{a\pi}{2}, \end{cases} \tag{3.23}$$

$$T(t) = 0, \qquad |t| \leqslant \frac{a\pi}{2}, \tag{3.24}$$

where P is a positive constant. Denote the segment $-l \leqslant t \leqslant l$ by γ_0, oriented from left to right.

In this case, by the results previously obtained, the stress functions are

$$\left.\begin{aligned} \Phi(z_1) &= \frac{i\nu_2 P}{2\pi(\nu_1 - \nu_2)}\left\{\ln\left|\sin\frac{t - z_1}{a}\right| + i\arg\sin\frac{t - z_1}{a}\right\}_{\gamma_0} - \frac{B_1 l P}{a\pi(G_1 - G_2)}, \\ \Psi(z_2) &= \frac{-i\nu_1 P}{2\pi(\nu_1 - \nu_2)}\left\{\ln\left|\sin\frac{t - z_2}{a}\right| + i\arg\sin\frac{t - z_2}{a}\right\}_{\gamma_0} - \frac{B_1 l P}{a\pi(G_1 - G_2)}. \end{aligned}\right\} \tag{3.25}$$

The stress distribution is:

$$\left.\begin{aligned} \sigma_x &= \frac{\nu_1\nu_2 P}{\pi(\nu_1 - \nu_2)}\left\{\nu_1\left[\arg\sin\frac{t - z_1}{a}\right]_{\gamma_0} - \nu_2\left[\arg\sin\frac{t - z_2}{a}\right]_{\gamma_0} + \frac{2B_1 l(\nu_1^2 - \nu_2^2)P}{a\pi(G_1 - G_2)}\right\}, \\ \sigma_y &= \frac{P}{\pi(\nu_1 - \nu_2)}\left\{-\nu_2\left[\arg\sin\frac{t - z_1}{a}\right]_{\gamma_0} + \nu_1\left[\arg\sin\frac{t - z_2}{a}\right]_{\gamma_0}\right\}, \\ \tau_{xy} &= \frac{\nu_1\nu_2 P}{\pi(\nu_1 - \nu_2)}\ln\left|\frac{\sin\dfrac{l - z_1}{a}\sin\dfrac{l + z_2}{a}}{\sin\dfrac{l + z_1}{a}\sin\dfrac{l - z_2}{a}}\right|. \end{aligned}\right\} \tag{3.26}$$

We point out that, if let $a \to \infty$ in (3.26), then we get stress distribution of the orthotropic half plane when a uniform pressure P is applied on a segment γ_0: $[-l, l]$ of its boundary:

$$\begin{aligned} \sigma_x &= \frac{\nu_1\nu_2 P}{\pi(\nu_1 - \nu_2)}\{\nu_1[\arg(t - z_1)]_{\gamma_0} - \nu_2[\arg(t - z_2)]_{\gamma_0}\}, \\ \sigma_y &= \frac{P}{\pi(\nu_1 - \nu_2)}\{-\nu_2[\arg(t - z_1)]_{\gamma_0} + \nu_2[\arg(t - z_2)]_{\gamma_0}\}, \\ \tau_{xy} &= \frac{\nu_1\nu_2 P}{\pi(\nu_1 - \nu_2)}\ln\left|\frac{(l - z_1)(l + z_2)}{(l + z_1)(l - z_2)}\right|. \end{aligned}$$

Let us then consider the case where there is a periodic set of concentrated pressure P_0 applied on the boundary. It is the case when $l \to 0$ and $P \to \infty$ with $2lP = P_0$ in (3.25) and (3.26). The obtained stress functions are

$$\left.\begin{aligned}\Phi(z_1) &= \frac{-i\nu_2 P_0}{2a\pi(\nu_1-\nu_2)}\cot\frac{z_1}{a} - \frac{B_1 P_0}{2a\pi(G_1-G_2)},\\ \Psi(z_2) &= \frac{i\nu_1 P_0}{2a\pi(\nu_1-\nu_2)}\cot\frac{z_2}{a} + \frac{B_1 P_0}{2a\pi(G_1-G_2)},\end{aligned}\right\} \tag{3.27}$$

while the stress distribution is given by

$$\left.\begin{aligned}\sigma_x &= -\frac{\nu_1\nu_2 P_0}{a\pi(\nu_1-\nu_2)}\left\{\nu_1 \mathrm{Im}[\cot\frac{z_1}{a}] - \nu_2\mathrm{Im}[\cot\frac{z_2}{a}]\right\} + \frac{B_1P_0}{a\pi(G_1-G_2)}(\nu_1^2-\nu_2^2),\\ \sigma_y &= \frac{P_0}{a\pi(\nu_1-\nu_2)}\left\{\nu_2\mathrm{Im}[\cot\frac{z_1}{a}] - \nu_1\mathrm{Im}[\cot\frac{z_2}{a}]\right\},\\ \tau_{xy} &= -\frac{\nu_1\nu_2P_0}{a\pi(\nu_1-\nu_2)}\left\{\mathrm{Re}[\cot\frac{z_1}{a}] - \mathrm{Re}[\cot\frac{z_2}{a}]\right\}.\end{aligned}\right\} \tag{3.28}$$

If let $a \to \infty$ in (3.28), then we get the stress distribution when the elastic half-plane S^- is subjected to a concentrated pressure force P_0:

$$\begin{aligned}\sigma_x &= -\frac{\nu_1\nu_2(\nu_1+\nu_2)P_0}{\pi}\frac{x^2y}{(x^2+\nu_1y^2)(x^2+\nu_2y^2)},\\ \sigma_y &= \frac{\nu_1\nu_2(\nu_1+\nu_2)P_0}{\pi}\frac{y^3}{(x^2+\nu_1y^2)(x^2+\nu_2y^2)},\\ \tau_{xy} &= -\frac{\nu_1\nu_2(\nu_1+\nu_2)P_0}{\pi}\frac{xy^2}{(x^2+\nu_1y^2)(x^2+\nu_2y^2)},\end{aligned}$$

which is identical to the result in Lekhnitsky [1].

Example 3.2 The elastic half-plane is the same as above, a uniform shearing force T is periodically applied to the segments as there, i.e.,

$$P(t) = 0, \quad |t| \leqslant \frac{a\pi}{2},$$

$$T(t) = \begin{cases} T, & |t| \leqslant l,\\ 0, & l < |t| \leqslant \frac{a\pi}{2},\end{cases}$$

where T is a constant.

In this case, the stress functions are

$$\left.\begin{aligned}\Phi(z_1) &= \frac{-T}{2(\nu_1-\nu_2)\pi}\left\{\ln\left|\sin\frac{t-z_1}{a}\right| + i\,\mathrm{argsin}\frac{t-z_1}{a}\right\}_{\gamma_0},\\ \Psi(z_2) &= \frac{T}{2(\nu_1-\nu_2)\pi}\left\{\ln\left|\sin\frac{t-z_2}{a}\right| + i\,\mathrm{argsin}\frac{t-z_2}{a}\right\}_{\gamma_0},\end{aligned}\right. \tag{3.29}$$

while the stress distribution is given by

$$\left.\begin{aligned}\sigma_x &= \frac{T}{\pi(\nu_1-\nu_2)}\left[\nu_1^2\ln\left|\frac{\sin\frac{l-z_1}{a}}{\sin\frac{l+z_1}{a}}\right| - \nu_2^2\ln\left|\frac{\sin\frac{l-z_2}{a}}{\sin\frac{l+z_2}{a}}\right|\right],\\ \sigma_y &= \frac{T}{\pi(\nu_1-\nu_2)}\ln\left|\frac{\sin\frac{l-z_2}{a}\sin\frac{l+z_1}{a}}{\sin\frac{l-z_1}{a}\sin\frac{l+z_2}{a}}\right|,\\ \tau_{xy} &= \frac{-T}{\pi(\nu_1-\nu_2)}\left[\nu_1\mathrm{argsin}\frac{l-z_1}{a} - \nu_2\mathrm{argsin}\frac{l-z_2}{a}\right]_{\gamma_0}.\end{aligned}\right\} \tag{3.30}$$

By letting $a\to\infty$ in (3.30), the stress distribution may be obtained when a periodic uniform shearing stress T is subjected on the segment γ: ($[-l, l]$ of the boundary:

$$\begin{aligned}\sigma_x &= \frac{T}{\pi(\nu_1-\nu_2)}\left[\nu_1^2\ln\left|\frac{l-z_1}{l+z_1}\right| - \nu_2^2\ln\left|\frac{l-z_2}{l+z_2}\right|\right],\\ \sigma_y &= \frac{T}{\pi(\nu_1-\nu_2)}\ln\left|\frac{(l-z_2)(l+z_1)}{(l-z_1)(l+z_2)}\right|,\\ \tau_{xy} &= \frac{-T}{\pi(\nu_1-\nu_2)}[\nu_1\arg|t-z_1| - \nu_2\arg|t-z_2|]_{\gamma_0}.\end{aligned}$$

Then, consider the case where periodic concentrated shearing force T_0 is is applied to the boundary (let $l\to 0$, $T\to+\infty$ with $2lT = T_0$ in (3.29)), we may get the stress functions

$$\left.\begin{aligned}\Phi(z_1) &= \frac{T_0}{2(\nu_1-\nu_2)\pi}\cot\frac{z_1}{a},\\ \Psi(z_2) &= \frac{-T_0}{2(\nu_1-\nu_2)\pi}\cot\frac{z_2}{a},\end{aligned}\right\} \tag{3.31}$$

while the stress distribution is given by

$$\left.\begin{aligned}\sigma_x &= \frac{T_0}{\pi(\nu_1-\nu_2)}\mathrm{Re}\left[-\nu_1^2\cot\frac{z_1}{a} + \nu_2^2\cot\frac{z_2}{a}\right],\\ \sigma_y &= \frac{T_0}{\pi(\nu_1-\nu_2)}\mathrm{Re}\left[\cot\frac{z_1}{a} - \cot\frac{z_2}{a}\right],\\ \tau_{xy} &= \frac{T_0}{\pi(\nu_1-\nu_2)}\mathrm{Im}\left[\nu_1\cot\frac{z_1}{a} - \nu_2\cot\frac{z_2}{a}\right].\end{aligned}\right\} \tag{3.32}$$

If let $a \to \infty$ in (3.32), then we have the stress distribution for the case where a single concentrated shearing force T_0 is applied at the origin:

$$\sigma_x = -\frac{(\nu_1+\nu_2)T_0}{\pi}\frac{x^3}{(x^2+\nu_1^2y^2)(x^2+\nu_2^2y^2)},$$
$$\sigma_y = -\frac{(\nu_1+\nu_2)T_0}{\pi}\frac{xy^2}{(x^2+\nu_1^2y^2)(x^2+\nu_2^2y^2)},$$
$$\tau_{xy} = \frac{(\nu_1+\nu_2)T_0}{\pi}\frac{x^2y}{(x^2+\nu_1^2y^2)(x^2+\nu_2^2y^2)}.$$

2. The second fundamental problem

Assume that, on the boundary x-axis of the anisotropic elastic half-plane S^-, the displacement

$$u^- + iv^- = g_1(t) + ig_2(t) \qquad (3.33)$$

is given, where $g_1(t)+ig_2(t)$ is continuous and $g_1'(t)+ig_2'(t)\in H$ arcwisely. Moreover, on the segment L_0 of the boundary in a period, the principal vector $X+iY$ of the external stresses is also given. Under these basic boundary conditions, find the equilibrium, called the *periodic second fundamental problem*.

Theorem 3.2 *Under the above assumptions, the solution of the second fundamental problem uniquely exists.*

Proof Applying (3.7) and (3.8), make the following linear combinations of u' and v':

$$\begin{aligned} q_2u' - p_2v' &= (p_1q_2-p_2q_1)\Phi(z_1) + (\bar{p}_1q_2-\bar{q}_1p_2)\overline{\Phi(z_1)} \\ &\quad + (q_2\bar{p}_2 - p_2\bar{q}_2)\overline{\Psi(z_2)}, \\ q_1u' - p_1v' &= (\bar{p}_1q_1-\bar{q}_1p_1)\overline{\Phi(z_1)} + (p_2q_1-q_2p_1)\Psi(z_2) \\ &\quad + (\bar{p}_2q_1-\bar{q}_2p_1)\overline{\Psi(z_2)}. \end{aligned}$$

Multiplying both sides of these two equations by $\frac{1}{2a\pi i}\cot\frac{t-z_1}{a}dt$ and $\frac{1}{2a\pi i}\cot\frac{t-z_2}{a}dt$ respectively and integrating along L_0, and noting (3.23), we have, by the formulas (1.59) and (1.60) for integrals with Hilbert kernel, we have

$$\Phi(z_1) = \frac{1}{q_1p_2-p_1q_2}\frac{1}{2a\pi i}\int_{L_0}(q_2g_1'-p_2g_2')\cot\frac{t-z_1}{a}dt + \gamma_1, \qquad (3.34)$$

$$\Psi(z_2) = \frac{-1}{q_1p_2-p_1q_2}\frac{1}{2a\pi i}\int_{L_0}(q_1g_1'-p_1g_2')\cot\frac{t-z_2}{a}dt + \gamma_2, \qquad (3.35)$$

where

$$\gamma_1 = -\frac{1}{2}\frac{1}{q_1p_2 - p_1q_2}[(p_1q_2 - q_1p_2)\Phi(-\infty i) - (\bar{p}_1q_2 - \bar{q}_1p_2)\overline{\Phi(-\infty i)}$$
$$- (q_2\bar{p}_2 - p_2\bar{q}_2)\overline{\Psi(-\infty i)}],$$
$$\gamma_2 = \frac{1}{2}\frac{1}{q_1p_2 - p_1q_2}[(p_2q_1 - q_2p_1)\Psi(-\infty i) - (\bar{p}_1q_1 - \bar{q}_1p_1)\overline{\Phi(-\infty i)}$$
$$- (\bar{p}_2q_1 - \bar{q}_2p_1)\overline{\Psi(-\infty i)}].$$

Evidently,

$$\mathrm{Re}[p_1\gamma_1 + p_2\gamma_2] = 0, \tag{3.36}$$

$$\mathrm{Re}[q_1\gamma_1 + q_2\gamma_2] = 0. \text{①} \tag{3.37}$$

In order to determine γ_1 and γ_2 completely, we need also consider the condition of equilibrium at $z = -\infty i$. By periodicity of $g_1'(t)$ and $g_2'(t)$, we have

$$\Phi(-\infty i) = \gamma_1, \quad \Psi(-\infty i) = \gamma_2.$$

By applying the expressions (3.3) and (3.4), the stresses at $z = -\infty i$ are

$$\sigma_y(-\infty i) = 2\mathrm{Re}[\Phi(-\infty i) + \Psi(-\infty i)],$$
$$\tau_{xy}(-\infty i) = -2\mathrm{Re}[\mu_1\Phi(-\infty i) + \mu_2\Psi(-\infty i)].$$

Then, we obtain another two equations in γ_1 and γ_2 because of (3.1):

$$\mathrm{Re}[\gamma_1 + \gamma_2] = \frac{Y}{2a\pi}, \tag{3.38}$$

$$\mathrm{Re}[\mu_1\gamma_1 + \mu_2\gamma_2] = -\frac{X}{2a\pi}. \tag{3.39}$$

Thus, γ_1 and γ_2 may be obtained by solving (3.36) – (3.39) and so the stress functions are completely determined. By substituting (3.34) and (3.35) into (3.2) – (3.4), the stress distribution in the whole elastic body for this problem may be achieved.

Let us illustrate a case usually occuring in practice.

When the elastic half-plane S^- is orthotropic as in the previous paragraph, the stress functions become

$$\Phi(z_1) = \frac{-1}{H_1G_2 - G_1H_2}\frac{1}{2a\pi}\int_{L_0}(iG_2g_1' - H_2g_2')\cot\frac{t - z_1}{a}dt + \gamma_1, \tag{3.40}$$

$$\Psi(z_2) = \frac{1}{H_1G_2 - G_1H_2}\frac{1}{2a\pi}\int_{L_0}(iG_1g_1' - H_1g_2')\cot\frac{t - z_2}{a}dt + \gamma_2, \tag{3.41}$$

where $\gamma_1 = \gamma_1^* + i\gamma_1'$ and $\gamma_2 = \gamma_2^* + i\gamma_2'$ are determined by the following system of linear equations:

---

① (3.36) and (3.37) may be deduced by periodicity of the displacements.

$$G_1\gamma_1^* + G_2\gamma_2^* = 0,\ H_1\gamma_1' + H_2\gamma_2' = 0,$$
$$\gamma_1^* + \gamma_2^* = \frac{Y}{2a\pi},\ \nu_1\gamma_1' + \nu_2\gamma_2' = \frac{X}{2a\pi},$$

and, as a result,

$$\left.\begin{aligned}\gamma_1^* &= -\frac{1}{2a\pi}\frac{G_2 Y}{G_1 - G_2},\quad \gamma_1' = -\frac{1}{2a\pi}\frac{H_2 X}{H_1\nu_2 - H_2\nu_1},\\ \gamma_2^* &= \frac{1}{2a\pi}\frac{G_1 Y}{G_1 - G_2},\quad \gamma_2' = \frac{1}{2a\pi}\frac{H_1 X}{H_1\nu_2 - H_2\nu_1}.\end{aligned}\right\} \tag{3.42}$$

Similar to the isotropic case, we also have the conclusion that $\sigma_x(-\infty i) = 0$ when and only when $Y = 0$.

Example 3.3 Consider the case where the given periodic displacements on the boundary are of wedge type, that is, in a period, the displacement on the line-segment L_0 of the boundary is given by

$$g_1(t) = 0,\ g_2(t) = \begin{cases}\varepsilon\left(\dfrac{|t|}{l} - 1\right), & |t| \leqslant l,\\ 0, & l < |t| < \dfrac{a\pi}{2},\end{cases}$$

as shown in Fig.2.3.

Besides, assume the principal vector of the external stresses on L_0 is $iY(X = 0,\ Y \neq 0)$. Find the equilibrium.

At this time,

$$g_1'(t) = 0,\ g_2'(t) = \begin{cases}-\varepsilon/l, & \text{when } t \in l_1: [-l, 0],\\ \varepsilon/l, & \text{when } t \in l_2: [0, l],\\ 0, & \text{when } t \in L_0 - l_1 - l_2.\end{cases}$$

Then, the stress functions become

$$\Phi(z_1) = \frac{-1}{H_1G_2 - G_1H_2}\frac{\varepsilon G_2}{2\pi l}\left\{i\left[\operatorname{argsin}\frac{t - z_1}{a}\right]_{l_1} - i\left[\operatorname{argsin}\frac{t - z_1}{a}\right]_{l_2}\right.$$
$$\left. + \ln\left|\frac{\sin^2\dfrac{z_1}{a}}{\sin\dfrac{l - z_1}{a}\sin\dfrac{l + z_1}{a}}\right|\right\} - \frac{1}{2a\pi}\frac{G_2 Y}{G_1 - G_2}, \tag{3.43}$$

$$\Psi(z_2) = \frac{1}{H_1G_2 - G_1H_2}\frac{\varepsilon G_1}{2\pi l}\left\{i\left[\operatorname{argsin}\frac{t - z_2}{a}\right]_{l_1} - i\left[\operatorname{argsin}\frac{t - z_2}{a}\right]_{l_2}\right.$$
$$\left. + \ln\left|\frac{\sin^2\dfrac{z_2}{a}}{\sin\dfrac{l - z_2}{a}\sin\dfrac{l + z_2}{a}}\right|\right\} + \frac{1}{2a\pi}\frac{G_1 Y}{G_1 - G_2}, \tag{3.44}$$

and the stresses are

$$\left.\begin{aligned}
\sigma_x &= \frac{1}{H_1G_2 - G_1H_2}\frac{\varepsilon}{l\pi}\left\{G_2\nu_1^2\ln\left|\frac{\sin^2\frac{z_1}{a}}{\sin\frac{l+z_1}{a}\sin\frac{l-z_1}{a}}\right|\right.\\
&\quad\left. - G_1\nu_2^2\ln\left|\frac{\sin^2\frac{z_2}{a}}{\sin\frac{l+z_2}{a}\sin\frac{l-z_2}{a}}\right|\right\} - \frac{1}{a\pi}\frac{Y}{G_1 - G_2}(G_2\nu_1^2 - G_1\nu_2^2),\\
\sigma_y &= \frac{-1}{H_1G_2 - G_1H_2}\frac{\varepsilon}{l\pi}\left\{G_2\ln\left|\frac{\sin^2\frac{z_1}{a}}{\sin\frac{l+z_1}{a}\sin\frac{l-z_1}{a}}\right|\right.\\
&\quad\left. - G_1\ln\left|\frac{\sin^2\frac{z_2}{a}}{\sin\frac{l+z_2}{a}\sin\frac{l-z_2}{a}}\right|\right\} + \frac{Y}{a\pi},\\
\tau_{xy} &= \frac{-1}{H_1G_2 - G_1H_2}\frac{\varepsilon}{l\pi}\left\{\nu_1G_2\left[\operatorname{argsin}\frac{t-z_1}{a}\right]_{\gamma_1} - \nu_1G_2\left[\operatorname{argsin}\frac{t-z_1}{a}\right]_{\gamma_2}\right.\\
&\quad\left. - \nu_2G_1\left[\operatorname{argsin}\frac{t-z_2}{a}\right]_{\gamma_1} + \nu_2G_1\left[\operatorname{argsin}\frac{t-z_2}{a}\right]_{\gamma_2}\right\}
\end{aligned}\right\} \tag{3.45}$$

If let $a \to \infty$ in (3.45), we obtain the stress distribution for a single displacement of wedge type on $[-l, l]$:

$$\left.\begin{aligned}
\sigma_x &= \frac{1}{H_1G_2 - G_1H_2}\frac{\varepsilon}{l\pi}\left[G_2\nu_1^2\ln\left|\frac{z_1^2}{l^2 - z_1^2}\right| - G_1\nu_2^2\ln\left|\frac{z_2^2}{l^2 - z_2^2}\right|\right],\\
\sigma_y &= \frac{-1}{H_1G_2 - G_1H_2}\frac{\varepsilon}{l\pi}\left[G_2\ln\left|\frac{z_1^2}{l^2 - z_1^2}\right| - G_1\ln\left|\frac{z_2^2}{l^2 - z_2^2}\right|\right],\\
\tau_{xy} &= \frac{-1}{H_1G_2 - G_1H_2}\frac{\varepsilon}{l\pi}\{\nu_1G_2[\arg(t - z_1)]_{\gamma_1} - \nu_1G_2[\arg(t - z_1)]_{\gamma_2}\\
&\quad - \nu_2G_1[\arg(t - z_2)]_{\gamma_1} + \nu_2G_1[\arg(t - z_2)]_{\gamma_2}\}.
\end{aligned}\right\} \tag{3.46}$$

§ 3. Periodic Contact Problem for Anisotropic Medium

In the present section, the contact problems for a periodic row of stamps pressed on the anisotropic half-plane S^- will be studied. Assume the coefficient of friction on the contact parts is ρ (>0). The case without friction may be discussed similarly or by letting $\rho = 0$.

1. Stress functions expressed in terms of boundary values of the stress components

To this purpose, consider the following linear combinations of σ_y and τ_{xy}:

$$\mu_1\sigma_y + \tau_{xy} = (\mu_1 - \bar{\mu}_1)\overline{\Phi(z_1)} + (\mu_1 - \mu_2)\Psi(z_2) + (\mu_1 - \bar{\mu}_2)\overline{\Psi(z_2)}, \tag{3.47}$$

$$\mu_2\sigma_y + \tau_{xy} = (\mu_2 - \mu_1)\Phi(z_1) + (\mu_2 - \bar{\mu}_1)\overline{\Phi(z_1)} + (\mu_2 - \bar{\mu}_2)\overline{\Psi(z_2)}. \tag{3.48}$$

Take the boundary values of these two equalities as z tends to points on L_0 from S^-. By Lemma 3.1, both $\Phi(z_1)$ and $\Psi(z_2)$ are periodic functions. Applying the formulas (1.59) and (1.60) for integrals with Hilbert kernel, we have

$$\Phi(z_1) = \frac{1}{2a\pi i(\mu_1 - \mu_2)}\int_{L_0}(\mu_2\sigma_y + \tau_{xy})_{y=0}\cot\frac{t - z_1}{a}dt, \tag{3.49}$$

$$\Psi(z_2) = -\frac{1}{2a\pi i(\mu_1 - \mu_2)}\int_{L_0}(\mu_1\sigma_y + \tau_{xy})_{y=0}\cot\frac{t - z_2}{a}dt, \tag{3.50}$$

Taking boundary values on both sides of (3.8), we have

$$\left(\frac{\partial v}{\partial x}\right)_{y=0} = [q_1\Phi(z_1) + \bar{q}_1\overline{\Phi(z_1)} + q_2\Psi(z_2) + \bar{q}_2\overline{\Psi(z_2)}]_{y=0}. \tag{3.51}$$

Substituting (3.49) and (3.50) into it, after simplification, we easily get

$$-\left(\frac{\partial v}{\partial x}\right)_{y=0} = \frac{A_3}{a\pi}\int_{L_0}(\sigma_y)_{y=0}\cot\frac{t - x}{a}dt + B_3(\sigma_y)_{y=0} + \frac{A_4}{a\pi}\int_{L_0}(\tau_{xy})_{y=0}\cot\frac{t - x}{a}dt + B_4(\tau_{xy})_{y=0}, \tag{3.52}$$

where

$$\left.\begin{aligned} A_3 &= \frac{\beta_{22}}{2i}\left(\frac{1}{\mu_1} - \frac{1}{\bar{\mu}_1} + \frac{1}{\mu_2} - \frac{1}{\bar{\mu}_2}\right), \\ B_3 &= -\frac{\beta_{22}}{2}\left(\frac{1}{\mu_1} + \frac{1}{\bar{\mu}_1} + \frac{1}{\mu_2} + \frac{1}{\bar{\mu}_2}\right) + \beta_{26}, \\ A_4 &= -\frac{\beta_{22}}{2i}\left(\frac{1}{\mu_1\mu_2} - \frac{1}{\bar{\mu}_1\bar{\mu}_2}\right), \\ B_4 &= -\frac{\beta_{22}}{2}\left(\frac{1}{\mu_1\mu_2} + \frac{1}{\bar{\mu}_1\bar{\mu}_2}\right) + \beta_{12}. \end{aligned}\right\} \tag{3.53}$$

By the way, we mention that (3.52) will play key role in the transformation of boundary conditions.

2. Formulation of the problem

Formulation of the periodic contact problem in anisotropic half-plane S^- is as follows.

Assume that a row of periodic stamps (with bases of the same shape) pressed on S^- and there exists friction between the stamps and S^- with coefficient of friction ρ, that is, beneath the stamps, the shearing stress $T(x) = \tau_{xy}(x)$ and normal pressure $P(x) = -\sigma_y(x)$ obey the Coulomb's law:

$$T(x) = \rho P(x),\ x \in \gamma_0$$

or

$$\tau_{xy}(x) + \rho\sigma_y(x) = 0,\ x \in \gamma_0,$$

where γ_0: $-l \leqslant x \leqslant l$, is the contact line-segment in L_0: $-\frac{1}{2}a\pi < x < \frac{1}{2}a\pi$. Assume the stamps are in the limiting situation of equilibrium. On the free interval $L_0 - \gamma_0$, there is no load, i.e.,

$$\sigma_y = 0,\ \tau_{xy} = 0.$$

Besides, assume the vertical displacement

$$v^-(\dot{x}) = f(x) \tag{3.54}$$

is given, where $y = f(x)$ is the equation of the bases of the stamps, which is $a\pi$-periodic with $f'(x) \in H$. Moreover, the external pressure force P_0 applied on each stamp is also given and so the principal vector of the external stresses is $X + iY = T_0 - iP_0 = (\rho - i)P_0$. Find the elastic equilibrium under these assumptions.

Thus, we have the boundary conditioins on L_0:

$$\left.\begin{aligned} &\sigma_y(x) = 0,\ \tau_{xy}(x) = 0,\ x \in L_0 - \gamma_0, \\ &\tau_{xy}(x) + \rho\sigma_y(x) = 0,\ v^-(x) = f(x),\ x \in \gamma_0. \end{aligned}\right\} \tag{3.55}$$

In order to solve the problem, we have to transform (3.55). Introduce two functions represented by integrals with Hilbert kernel:

$$w_1(z) = u_1 - iv_1 = \int_{L_0} \sigma_y(t)\cot\frac{t-z}{a}dt, \tag{3.56}$$

$$w_2(z) = u_2 - iv_2 = \int_{L_0} \tau_{xy}(t)\cot\frac{t-z}{a}dt + \beta, \tag{3.57}$$

where β is an undetermined real constant.①

As z tends to points L_0 from S^-, we have, by the generalized Plemelj formula (1.14),

$$w_1^-(x) = u_1^- - iv_1^- = \int_{L_0} \sigma_y(t)\cot\frac{t-x}{a}dt - a\pi i\sigma_y(x),$$

① The necessity of occurence of β will be clear in paragraph 4 below.

$$w_2^-(x) = u_2^- - iv_2^- = \int_{L_0} \tau_{xy}(t)\cot\frac{t-x}{a}dt - a\pi i\tau_{xy}(x) + \beta.$$

On seperating their real and imaginary parts, the following equations are obtained:

$$\left.\begin{aligned} \sigma_y(x) &= \frac{1}{a\pi}v_1^-(x) = \frac{1}{a\pi}\mathrm{Im}\,w_1^-(x), \\ \tau_{xy}(x) &= \frac{1}{a\pi}v_2^-(x) = \frac{1}{a\pi}\mathrm{Im}\,w_2^-(x), \\ u_1^-(x) &= \int_{L_0}\sigma_y(t)\cot\frac{t-x}{a}dt, \\ u_2^-(x) &= \int_{L_0}\tau_{xy}(t)\cot\frac{t-x}{a}dt + \beta. \end{aligned}\right\} \qquad (G)$$

Since there does not occur concentrated stresses on the bases of the stamps (including their tips), functions $w_1(z)$ and $w_2(z)$, as represented by integrals with Hilbert kernel in (3.56) and (3.57) respectively, could have integrable singularity at any tip x_0 of the stamps of the form $(\tan\frac{z}{a} - \tan\frac{x_0}{a})^{-\lambda}$, $0<\lambda<1$.

On account of $v^-(x) = f(x)$, (3.52) becomes

$$\begin{aligned} -f'(x) = {} & \frac{A_3}{a\pi}\int_{\gamma_0}\sigma_y(t)\cot\frac{t-x}{a}dt + \frac{B_3}{a\pi}\sigma_y(x) \\ & + \frac{A_4}{a\pi}\int_{\gamma_0}\tau_{xy}(t)\cot\frac{t-x}{a}dt + \frac{B_4}{a\pi}\tau_{xy}(x), \qquad x \in \gamma_0. \end{aligned} \qquad (3.58)$$

Consequently, we may transform condition (3.55) to, by applying the boundary values of
$w_1(z)$ and $w_2(z)$,

$$\left.\begin{aligned} & v_1^-(x) = 0, \ v_2^-(x) = 0, \ x \in L_0 - \gamma_0, \\ & -f'(x) = \frac{A_3}{a\pi}u_1^-(x) + \frac{B_3}{a\pi}v_1^-(x) + \frac{A_4}{a\pi}u_2^-(x) + \frac{B_4}{a\pi}v_2^-(x), \ x \in \gamma_0, \\ & v_2^-(x) + \rho v_1^-(x) = 0, \ x \in \gamma_0. \end{aligned}\right\} \qquad (3.59)$$

Thus, by noting the last equation, the following boundary conditions will be fulfilled by $w_1(z)$:

$$\left.\begin{aligned} & v_1^-(x) = 0, \ x \in L_0 - \gamma_0, \\ & u_1^-(x) + \frac{B_3 - \rho B_4}{A_3 - \rho A_4}v_1^-(x) = -\frac{a\pi}{A_3 - \rho A_4}f'(x), \ x \in \gamma_0. \end{aligned}\right\} \qquad (3.60)$$

In fact, this boundary value problem is: find a periodic function $w_1(z) = u_1 - iv_1$, holomorphic in S^- and satisfying on L_0

$$a(x)u_1^-(x) + b(x)v_1^-(x) = F(x),\ x \in L_0, \tag{3.61}$$

where

$$a(x) = \begin{cases} 1, & x \in \gamma_0; \\ 0, & x \in \gamma_0' = L_0 - \gamma_0, \end{cases} \qquad b(x) = \begin{cases} \dfrac{B_3 - \rho B_4}{A_3 - \rho A_4}, & x \in \gamma_0, \\ 1, & x \in \gamma_0'; \end{cases} \tag{3.62}$$

$$F(x) = \begin{cases} -\dfrac{a\pi}{A_3 - \rho A_4} f'(x), & x \in \gamma_0, \\ 0, & x \in \gamma_0'; \end{cases} \tag{3.63}$$

and the same for $x \in \gamma$ and γ' (γ, γ' being the unions of the segments congruent to γ_0 and γ_0' respectively). This is a particular periodic Riemann-Hilbert boundary value problem of the half-plane S^- discussed in § 2.2, Chapter I.

We shall give its solution in the next paragraph.

3. Solution of the problem

We note that $a(x)$, $b(x)$ and $F(x)$ have discontinuities at $x = \pm l$ ($|l| < \frac{1}{2}a\pi$) and so $w_1(z)$ has singularities at these points on L_0.

According to (1.38), in our case, we have $a_1 + ib_1 = 1 + i\dfrac{B_3 - \rho B_4}{A_3 - \rho A_4}$, $a_2 + ib_2 = i$ and hence, by (1.39),

$$\omega(x) = \begin{cases} \omega_1 = \arctan \dfrac{B_3 - \rho B_4}{A_3 - \rho A_4}, & x \in \gamma, \\ \omega_2 = \dfrac{\pi}{2}, & x \in \gamma', \end{cases} \tag{3.64}$$

where we have taken $0 < |\omega_1| < \frac{\pi}{2}$. Therefore, in this case, by (1.51),

$$\nu = \frac{\omega_2 - \omega_1}{\pi} = \frac{1}{2} - \theta,\ \theta = \frac{\omega_1}{\pi},\ 0 < \nu < 1. \tag{3.65}$$

Thus, by (1.53),

$$X(z) = \pm ie^{\pm \omega_1 i} E(z),\ z \in S^\pm, \tag{3.66}$$

where

$$E(z) = \left(\tan\frac{l}{a} - \tan\frac{z}{a}\right)^{-\frac{1}{2}+\theta}\left(\tan\frac{l}{a} + \tan\frac{z}{a}\right)^{-\frac{1}{2}-\theta},\ \operatorname{Im} z \neq 0, \tag{3.67}$$

is a holomorphic function in the z-plane cut by γ', which takes non-negative values when $z = t \in \gamma$, and so

$$X^{+}(t) = ie^{\omega_1 i}E(t), \; t \in \gamma, \tag{3.68}$$

where

$$E(t) = (\tan\frac{l}{a} - \tan\frac{t}{a})^{-\frac{1}{2}+\theta}(\tan\frac{l}{a} + \tan\frac{t}{a})^{-\frac{1}{2}-\theta}, \; t \in \gamma, \tag{3.69}$$

is positive, i.e., both the radical factors on its right side are positive on γ.

Therefore, by (1.54), the corresponding

$$\Omega_0(z) = \pm\, ie^{\pm\omega_1 i}(C_1\tan\frac{z}{a} + C_2)E(z), \; z \in S^{\pm}, \tag{3.70}$$

where C_1 and C_2 are undetermined real constants, and, on account of

$$a_1 - ib_1 = \sqrt{a_1^2 + b_1^2}e^{-\omega_1 i} = \frac{e^{-\omega_1 i}}{\cos\omega_1},$$

by (1.56),

$$\Omega_1(z) = \frac{\pm\, ie^{\pm\omega_1 i}\cos\omega_1 E(z)}{A_3 - \rho A_4}\int_{\gamma_0}\frac{f'(t)}{E(t)}\cot\frac{t - z}{a}dt, \; z \in S^{\pm}. \tag{3.71}$$

Therefore, we obtain

$$w_1(z) = \frac{\pm\, ie^{\pm\theta\pi i}\cos\theta\pi E(z)}{A_3 - \rho A_4}\int_{\gamma_0}\frac{f'(t)}{E(t)}\cot\frac{t - z}{a}$$
$$\pm\, ie^{\pm\omega_1 i}(C_1\tan\frac{z}{a} + C_2)E(z), \; z \in S^{\pm}, \tag{3.72}$$

and our solution is $w_1(z)$ when $z \in S^{-}$.

As for $w_2(z)$, by boundary condition, it is evident that

$$w_2(z) = -\rho w_1(z) + \beta. \tag{3.73}$$

4. Periodic condition for displacements and equilibrium condition at $z = -\infty i$

To obtain our solution completely, the real constants C_1, C_2 and β ought to be determined. To this aim, the periodic condition of the displacements and the condition of elastic
equilibrium at $z = -\infty i$ should be considered.

First, consider the former. We have the expression for the displacement function

$$u + iv = (p_1 + iq_1)\varphi(z_1) + (\bar{p}_1 + i\bar{q}_1)\overline{\varphi(z_1)} + (p_2 + iq_2)\psi(z_2)$$
$$+ (\bar{p}_2 + i\bar{q}_2)\overline{\psi(z_2)}.$$

Taking increments on its both sides along Λ_- (see Fig.2.6), we get

$$[u + iv]_{\Lambda_-} = (p_1 + iq_1)\int_{\Lambda_-} \Phi(z_1)dz_1 + (\bar{p}_1 + i\bar{q}_1)\int_{\Lambda_-} \overline{\Phi(z_1)}dz_1$$
$$+ (p_2 + iq_2)\int_{\Lambda_-} \Psi(z_1)dz_1 + (\bar{p}_2 + i\bar{q}_2)\int_{\Lambda_-} \overline{\Psi(z_2)}dz_2. \qquad (3.74)$$

By using (3.49) and (3.50), $\Phi(z_1)$ and $\Psi(z_2)$ may be expressed in terms of $w_1(z)$ and
$w_2(z)$:

$$\left.\begin{aligned} \Phi(z_1) &= \frac{1}{2a\pi i(\mu_1 - \mu_2)}[\mu_2 w_1(z_1) + w_2(z_1)], \\ \Psi(z_2) &= \frac{-1}{2a\pi i(\mu_1 - \mu_2)}[\mu_1 w_1(z_2) + w_2(z_2)], \end{aligned}\right. \qquad (3.75)$$

and by substitution into (3.74),

$$[u + iv]_{\Lambda_-} = \frac{1}{2a\pi i(\mu_1 - \mu_2)}\left\{[\mu_2(p_1 + iq_1) - \mu_1(p_2 + iq_2)]\int_{\Lambda_-} w_1(z)dz\right.$$
$$\left. + [(p_1 + iq_1) - (p_2 + iq_2)]\int_{\Lambda_-} w_2(z)dz\right\}$$
$$- \frac{1}{2a\pi i(\bar{\mu}_1 - \bar{\mu}_2)}\left\{[\bar{\mu}_2(\bar{p}_1 + i\bar{q}_1) - \bar{\mu}_1(\bar{p}_2 + i\bar{q}_2)]\int_{\Lambda_-} \overline{w_1(z)}dz\right.$$
$$\left. + [(\bar{p}_1 + i\bar{q}_1) - (\bar{p}_2 + i\bar{q}_2)]\int_{\Lambda_-} \overline{w_2(z)dz}\right\},$$

which must be equal to zero by the periodic condition of displacements. Thus, it becomes, on account of (3.73),

$$(\varepsilon_1 + i\delta_1)\int_{\Lambda_-} w_1(z)dz + (\varepsilon_2 + i\delta_2)\int_{\Lambda_-} \overline{w_1(z)}dz + (\varepsilon_3 + i\delta_3)\beta a\pi = 0, \qquad (3.76)$$

where we have put

$$\left.\begin{aligned} \varepsilon_1 + i\delta_1 &= \frac{1}{\mu_1 - \mu_2}\{[\mu_2(p_1 + iq_1) - \mu_1(p_2 + iq_2)] \\ &\quad - \rho[(p_1 + iq_1) - (p_2 + iq_2)]\}, \\ \varepsilon_2 + i\delta_2 &= -\frac{1}{\bar{\mu}_1 - \bar{\mu}_2}\{[\bar{\mu}_2(\bar{p}_1 + i\bar{q}_1) - \bar{\mu}_1(\bar{p}_2 + i\bar{q}_2)] \\ &\quad - \rho[(\bar{p}_1 + i\bar{q}_1) - (\bar{p}_2 + i\bar{q}_2)]\}, \\ \varepsilon_3 + i\delta_3 &= \frac{1}{\mu_1 - \mu_2}[(p_1 + iq_1) - (p_2 + iq_2)] \\ &\quad - \frac{1}{\bar{\mu}_1 - \bar{\mu}_2}[(\bar{p}_1 + i\bar{q}_1) - (\bar{p}_2 + i\bar{q}_2)]. \end{aligned}\right\} \qquad (3.77)$$

To evaluate the two integrals appeared in (3.76), we note that

$$\left.\begin{aligned}
&E(-\infty i) = -e^{i\theta\pi}\cos\frac{l}{a}e^{-i\frac{2l\theta}{a}},\\
&E(+\infty i) = e^{-i\theta\pi}\cos\frac{l}{a}e^{i\frac{2l\theta}{a}},\\
&\int_{\Lambda_\pm} E(z)dz = a\pi E(\pm\infty i) = \pm a\pi e^{-i\theta\pi}\cos\frac{l}{a}e^{\pm i\frac{2l\theta}{a}},\\
&\int_{\Lambda_\pm} E(z)\tan\frac{z}{a}dz = \int_{\Lambda_\pm} E(z)\frac{t-z}{a}dz = \pm a\pi i e^{-i\theta\pi}\cos\frac{l}{a}e^{\pm i\frac{2l\theta}{a}}.
\end{aligned}\right\}\tag{3.78}$$

By substitution into (3.72), we obtain

$$\int_{\Lambda_-} w(z)dz = \frac{-a\pi\cos\frac{l}{a}\cos\theta\pi e^{-i\frac{2l\theta}{a}}}{A_3-\rho A_4}\int_{\gamma_0}\frac{f'(t)}{E(t)}dt + a\pi\cos\frac{l}{a}e^{-i\frac{2l\theta}{a}}C_1 + ia\pi\cos\frac{l}{a}e^{-i\frac{2l\theta}{a}}C_2.$$

Since $\overline{w_1(z)} = w_1(\bar{z})$ and $dz = d\bar{z}$ on Λ_-, we have

$$\begin{aligned}
\int_{\Lambda_-}\overline{w_1(z)}dz &= \int_{\Lambda_-} w_1(\bar{z})dz = \int_{\Lambda_-} w_1(\bar{z})d\bar{z} = \overline{\int_{\Lambda_-} w_1(z)dz}\\
&= \frac{-a\pi\cos\frac{l}{a}\cos\theta\pi e^{i\frac{2l\theta}{a}}}{A_3-\rho A_4}\int_{\gamma_0}\frac{f'(t)}{E(t)}dt\\
&\quad + a\pi\cos\frac{l}{a}e^{i\frac{2l\theta}{a}}C_1 - ia\pi\cos\frac{l}{a}e^{i\frac{2l\theta}{a}}C_2.
\end{aligned}$$

Thus, the condition (3.76) of periodicity of the displacements reduces to

$$\begin{aligned}
&\{(\varepsilon_1+i\delta_1)e^{-i\frac{2l\theta}{a}} + (\varepsilon_2+i\delta_2)e^{i\frac{2l\theta}{a}}\}C_1\\
&+ i\{(\varepsilon_1+i\delta_1)e^{-i\frac{2l\theta}{a}} - (\varepsilon_2+i\delta_2)e^{i\frac{2l\theta}{a}}\}C_2 + (\varepsilon_3+i\delta_3)\beta a\pi\\
= &\frac{\cos\pi\theta}{A_3-\rho A_4}\{(\varepsilon_1+i\delta_1)e^{-i\frac{2l\theta}{a}} + (\varepsilon_2+i\delta_2)e^{-i\frac{2l\theta}{a}}\}\int_{\gamma_0}\frac{f'(t)}{E(t)}dt.
\end{aligned}\tag{3.79}$$

Next, consider the condition of equilibrium at $z = -\infty i$, which is, by (3.1),

$$\sigma_y(-\infty i) = -\frac{P_0}{a\pi}.\tag{3.80}$$

By substituting (3.75) into (3.3), noting (3.73) and letting $z_1, z_2 \to -\infty i$, condition (3.80) is transferred to

$$P_0 = \operatorname{Im} w_1(-\infty i).\tag{3.81}$$

Let $z \to -\infty i$ in (3.72). By the first equality in (3.78), we have

$$w_1(-\infty i) = \frac{-\cos\frac{l}{a}\cos\pi\theta e^{-i\frac{2l\theta}{a}}}{A_3 - \rho A_4}\int_{\gamma_0}\frac{f'(t)}{E(t)}dt + (C_1 + iC_2)\cos\frac{l}{a}e^{-i\frac{2l\theta}{a}}.$$

Substituting it into (3.81), condition (3.80) becomes

$$\frac{P_0}{\cos\frac{l}{a}} - \frac{\cos\pi\theta\sin\frac{2l\theta}{a}}{A_3 - \rho A_4}\int_{\gamma_0}\frac{f'(t)}{E(t)}dt = C_2\cos\frac{2l\theta}{a} - C_1\sin\frac{2l\theta}{a}. \quad (3.82)$$

Thus, C_1, C_2 and β may be usiquely determined by (3.79) (two real equations indeed) and (3.82).

5. The pressure beneath the stamps

In order to find the pressure distribution $p(x)$ applied on γ_0, we note that $(\sigma_y)_{y=0} = -p(x)$, $x \in \gamma_0$, and, by the first equation in (G), we have

$$p(x) = -\frac{1}{a\pi}\text{Im}[w_1^-(x)],\ x \in \gamma_0.$$

By the generalized Plemelj formula (1.14), we get

$$p(x) = \frac{1}{2(A_3 - \rho A_4)}\left\{-\sin 2\theta\pi f'(x) + \frac{2\cos^2\theta\pi E(x)}{a\pi}\int_{\gamma_0}\frac{f'(t)}{E(t)}\cot\frac{t-x}{a}dt\right\}$$
$$+ \frac{\cos\theta\pi}{a\pi}E(x)(C_1\tan\frac{x}{a} + C_2),\ x \in \gamma_0. \quad (3.83)$$

Example 3.4 Consider the case where the stamps possess periodic horizontal rectilinear bases.

Here, $f'(x) = 0$. Then, by (3.83),

$$p(x) = \frac{\cos\theta\pi}{a\pi}E(x)(C_1\tan\frac{x}{a} + C_2),\ x \in \gamma_0, \quad (3.84)$$

where C_1 and C_2 are determined by

$$\{(\varepsilon_1 + i\delta_1)e^{-i\frac{2l\theta}{a}} + (\varepsilon_2 + i\delta_2)e^{i\frac{2l\theta}{a}}\}C_1 + i\{(\varepsilon_1 + i\delta_1)e^{-i\frac{2l\theta}{a}}$$
$$- (\varepsilon_2 + i\delta_2)e^{i\frac{2l\theta}{a}}\}C_2 + (\varepsilon_3 + i\delta_3)a\pi\beta = 0,$$
$$C_2\cos\frac{2l\theta}{a} - C_1\sin\frac{2l\theta}{a} = \frac{P_0}{\cos\frac{l}{a}}$$

(see (3.79), (3.82)).

Chapter IV

Problems with Periodic Moving Loads on Isotropic Elastic Half-Plane

§ 1. Stress Functions and Fundamental Problems

The problems with periodic moving loads on the boundary of an isotropic half-plane will be studied in the present chapter, which is, in some respect, similar to the problems considered in the previous chapter. We will discuss the periodicity of the stress functions, fundamental problems and contact problems.

1. Periodicity of the stress functions for isotropic medium with moving loads

It is known that, when the loads move with constant velocity w along the boundary of an isotropic half-plane S^-, the stress components σ_x, σ_y, τ_{xy} and the displacement components u, v may be expressed in terms of two complex stress functions $\varphi(z_1)$ and $\psi(z_2)$ by

$$\sigma_y = i[K\Phi(z_1) - K\overline{\Phi(z_1)} + F\Psi(z_2) - F\overline{\Psi(z_2)}], \tag{4.1}$$
$$\sigma_y = i[G\Phi(z_1) - G\overline{\Phi(z_1)} + H\Psi(z_2) - H\overline{\Psi(z_2)}], \tag{4.2}$$
$$\tau_{xy} = M\Phi(z_1) + M\overline{\Phi(z_1)} + N\Psi(z_2) + N\overline{\Psi(z_2)}, \tag{4.3}$$
$$u = i[A\varphi(z_1) - A\overline{\varphi(z_1)} + B\psi(z_2) - B\overline{\psi(z_2)}], \tag{4.4}$$
$$v = C\varphi(z_1) + C\overline{\varphi(z_1)} + D\psi(z_2) + D\overline{\psi(z_2)}, \tag{4.5}$$

where $\varphi(z_1)$ and $\psi(z_2)$ are functions holomorphic in $z_1 = x + ik_1y$ and $z_2 = x + ik_2y$ respectively for $z = x + iy \in S^-$ and $\Phi(z_1) = \varphi'(z_1)$, $\Psi(z_2) = \psi'(z_2)$, in which

$$A = ak_1,\ B = ak_2,\ C = b - k_1^2,\ D = b - k_2^2,$$
$$K = ck_1 - dk_1^3,\ F = ck_2 - dk_2^3,$$
$$G = fk_1 - gk_1^3,\ H = fk_2 - gk_2^3,$$

$$M = -hk_1^2 + l, \quad N = -hk_2^2 + l,$$

and

$$k_1^2 = 1 - \frac{w^2}{c_1}, \quad k_2^2 = 1 - \frac{w^2}{c_2},$$

$$a = -\frac{\lambda + \mu}{\mu}, \quad b = \frac{\lambda + 2\mu}{\mu} - \frac{w^2}{c_2^2},$$

$$c = -(\lambda + 2\mu) - \frac{\lambda w^2}{c_1^2}, \quad d = \lambda,$$

$$f = (3\lambda + 4\mu) - (\lambda + 2\mu)\frac{w^2}{c_2^2}, \quad g = \lambda + 2\mu,$$

$$l = \lambda + 2\mu - \frac{\mu w^2}{c_2^2}, \quad h = -\lambda,$$

while

$$c_1 = \sqrt{\frac{\lambda + 2\mu}{\delta}}, \quad c_2 = \sqrt{\frac{\mu}{\delta}}$$

are the propagation velocities of the expansion wave and the distorsion wave respectively, ① and as usual, δ denotes the mass of unit volume of the medium and λ, μ, the Lamé constants (Cf. Galin [1]).

Lemma 4.1 *Under the basic assumptions, the stress functions* $\varphi'(z_1) = \Phi(z_1)$ *are* $\psi'(z_2) = \Psi(z_2)$ *are* $a\pi$*-periodic holomorphic functions.*

The proof of this lemma is similar to that of Lemma 3.1, which will be omitted here.

We see that the situation considered here is similar to the anisotropic elastic statics in certain respect.

2. Formulation and solution of the problems

Assume the isotropic elastic medium occupys the lower half-plane S^-. Let $z = \tau$ (real) is an arbitrary point on its boundary x-axis. Assume the periodic normal and tangential loads on its boundary

$$\sigma_y(\tau) = -P(\tau), \quad \tau_{xy}(\tau) = T(\tau) \quad (\tau\text{: real}) \tag{4.6}$$

($P(\tau)$, $T(\tau) \in H$ arcwisely) are moving with constant velocity w. Find the dynamic elastic equilibrium.

According to (4.2) and (4.3),

$$N\sigma_y - iH\tau_{xy} = i(GN - MH)\Phi(z_1) - i(GN + MH)\overline{\Phi(z_1)} - 2iHN\overline{\Psi(z_2)}.$$

---

① We use, as in literature, the letter "a" to denote the quantity defined here which is appeared in A, B and l, not that in the "period $a\pi$" which will be always used in the sequel.

Then we have, as z_1, z_2 tend to τ on the x-axis, by (4.6),

$$-NP(\tau) - iHT(\tau) = i(GN - MH)\Phi(\tau) - i(GN + MH)\overline{\Phi(\tau)} - 2iH\overline{\Psi(\tau)} \tag{4.7}$$

($\Phi(\tau) = \Phi^-(\tau)$, $\Psi(\tau) = \Psi^-(\tau)$). Similarly,

$$-MP(\tau) - iGT(\tau) = i(MH - GN)\Psi(\tau) - i(MH + GN)\overline{\Psi(\tau)} - 2iGM\overline{\Phi(\tau)}. \tag{4.8}$$

Multiplying both sides of (4.7) and (4.8) by $\frac{1}{2a\pi i}\cot\frac{\tau - z_1}{a}$ and $\frac{1}{2a\pi i}\cot\frac{\tau - z_2}{a}$ respectively and integating along L_0: $(-\frac{a\pi}{2}, \frac{a\pi}{2})$, by (1.59) and (1.60), we obtain the integral expressions of the stress functions:

$$\Phi(z_1) = \frac{-1}{2a\pi(GN - MH)}\int_{L_0}[NP(\tau) + iHT(\tau)]\cot\frac{\tau - z_1}{a}d\tau + \gamma_1, \tag{4.9}$$

$$\Psi(z_2) = \frac{1}{2a\pi(GN - MH)}\int_{L_0}[NP(\tau) + iHT(\tau)]\cot\frac{\tau - z_2}{a}d\tau + \gamma_2, \tag{4.10}$$

where

$$\gamma_1 = \frac{1}{2(GN - MH)}\{(GN - MH)\Phi(-\infty i) + (GN + MH)\overline{\Phi(-\infty i)} + 2HN\overline{\Psi(-\infty i)}\},$$

$$\gamma_2 = \frac{-1}{2(GN - MH)}\{(MH - GN)\Phi(-\infty i) + (MH + GN)\overline{\Psi(-\infty i)} + 2GM\overline{\Phi(-\infty i)}\}.$$

3. Periodic condition for displacements and equilibrium condition at $z = -\infty i$

We would use the periodic condition for displacements:

$$[u + iv]_{\Lambda_-} = 0, \tag{*}$$

where Λ_- is shown in Fig.2.6, and the equilibrium condition at $z = -\infty i$:

$$\sigma_y(-\infty i) = -\frac{1}{a\pi}\int_{L_0}P(\tau)d\tau = -P^*,$$
$$\tau_{xy}(-\infty i) = \frac{1}{a\pi}\int_{L_0}T(\tau)d\tau = T^*, \tag{**}$$

to determining γ_1 and γ_2.

First, consider the former. By integrating (4.9) and (4.10) respectively along L_0, we have, up to a rigid translation of the elastic body, we have

$$\varphi(z_1) = \frac{-1}{2\pi(GN - MH)}\int_{L_0}[NP(\tau) + iHT(\tau)]\text{logsin}\frac{\tau - z_1}{a}d\tau + \gamma_1 z_1,$$
$$\psi(z_2) = \frac{1}{2\pi(GN - MH)}\int_{L_0}[MP(\tau) + iGT(\tau)]\text{logsin}\frac{\tau - z_2}{a}d\tau + \gamma_2 z_2,$$

and so

$$\varphi(z_1 + a\pi) - \varphi(z_1) = \frac{-iNP^* + HT^*}{2(GN - MH)} + a\pi\gamma_1,$$
$$\psi(z_2 + a\pi) - \psi(z_2) = \frac{iMP^* - GT^*}{2(GN - MH)} + a\pi\gamma_2.$$

By substitution into (*), the following two equations are obtained, by (4.4) and (4.5),

$$A\,\text{Im}\gamma_1 + B\,\text{Im}\gamma_2 = \frac{AN - BM}{2a\pi(GN - MH)}P^*, \tag{4.11}$$
$$C\text{Re}\gamma_1 + D\text{Re}\gamma_2 = \frac{DG - CH}{2a\pi(GN - MH)}T^*. \tag{4.12}$$

Next, consider the equilibrium condition at $z = -\infty i$. By substituing

$$\Phi(-\infty i) = \frac{-1}{2a\pi i(GN - MH)}(NP^* + iHT^*) + \gamma_1,$$
$$\Psi(-\infty i) = \frac{1}{2a\pi i(GN - MH)}(NP^* + iGT^*) + \gamma_2$$

into (**), another two equations are obtained, by (4.2) and (4.3),

$$G\,\text{Im}\gamma_1 + H\,\text{Im}\gamma_2 = 0, \tag{4.13}$$
$$M\text{Re}\gamma_1 + N\text{Re}\gamma_2 = 0. \tag{4.14}$$

Thus, γ_1 and γ_2 are uniquely determined by (4.11) – (4.14). Therefore, the solution of our problem is obtained and, in the mean time, the unique existence of the solution is established.

4. A special case

We assume that, a uniform pressure, periodically districbuted and moving with constant velecity w, is applied to partial boundary of the elastic half-plane S^-: on the segment L_0 of a period of the boundary:

$$P(\tau) = \begin{cases} P, & |\tau| \leqslant l, \\ 0, & l < \tau \leqslant \frac{a\pi}{2}; \end{cases} \quad T(t) = 0, \quad |\tau| \leqslant \frac{a\pi}{2},$$

where P is a positive constant. Find the dynamic elastic equilibrium.

In this case, by (4.11) – (4.14),

$$\text{Re}\gamma_1 = \text{Re}\gamma_2 = 0,$$

$$\operatorname{Im}\gamma_1 = \frac{H(MB - AN)lP}{a\pi(BG - AH)(GN - MH)},$$
$$\operatorname{Im}\gamma_2 = \frac{-G(MB - AN)lP}{a\pi(BG - AH)(GN - MH)}.$$

Thus, the stress functions are expressed by

$$\Phi(z_1) = \frac{-NP}{2a\pi(GN - MH)}\left[\ln\left|\sin\frac{\tau - z_1}{a}\right| + i\operatorname{argsin}\frac{\tau - z_1}{a}\right]_{\gamma_0} + \frac{iH(MB - AN)lP}{a\pi(BG - AH)(GN - MH)},$$
$$\Psi(z_2) = \frac{MP}{2a\pi(GN - MH)}\left[\ln\left|\sin\frac{\tau - z_2}{a}\right| + i\operatorname{argsin}\frac{\tau - z_2}{a}\right]_{\gamma_0} - \frac{iG(BM - AN)lP}{a\pi(BG - AH)(GN - MH)},$$

where γ_0: $[-l, l]$, and the stress distribution, by

$$\sigma_x = \frac{-P}{a\pi(GN - MH)}\left\{KN\left[\operatorname{argsin}\frac{\tau - z_1}{a}\right]_{\gamma_0} - FM\left[\operatorname{argsin}\frac{\tau - z_2}{a}\right]_{\gamma_0}\right\} - \frac{2Pl(MB - AN)(KH - FG)}{a\pi(BG - AH)(GN - MH)},$$
$$\sigma_y = \frac{-P}{a\pi(GN - MH)}\left\{GN\left[\operatorname{argsin}\frac{\tau - z_1}{a}\right]_{\gamma_0} - HM\left[\operatorname{argsin}\frac{\tau - z_2}{a}\right]_{\gamma_0}\right\},$$
$$\tau_{xy} = \frac{-PMN}{a\pi(GN - MH)}\ln\left|\frac{\sin\dfrac{l - z_1}{a}\sin\dfrac{l + z_2}{a}}{\sin\dfrac{l + z_1}{a}\sin\dfrac{l - z_2}{a}}\right|.$$

If periodic concentrated loads moving with constant velocity w are applied on the boundary of S^-, i. e., as $l \to 0$, $P \to +\infty$ with $2lP = P_0$ (a finite constant), then, the stress functions become

$$\Phi(z_1) = \frac{NP_0}{2a\pi(GN - MH)}\cot\frac{z_1}{a} + \frac{iH(MB - AN)P_0}{2a\pi(BG - AH)(GN - MH)},$$
$$\Psi(z_2) = \frac{-MP_0}{2a\pi(GN - MH)}\cot\frac{z_2}{a} - \frac{iG(MB - AN)P_0}{2a\pi(BG - AH)(GN - MH)},$$

and the stress distribution is

$$\sigma_x = \frac{P_0}{a\pi(GN - MH)}\left\{KN\operatorname{Im}\left[\cot\frac{z_1}{a}\right] - FM\operatorname{Im}\left[\cot\frac{z_2}{a}\right]\right\} - \frac{P_0(MB - AN)(KH - FG)}{a\pi(BG - AH)(GN - MH)},$$
$$\sigma_y = \frac{-P_0}{a\pi(GN - MH)}\left\{GN\operatorname{Im}\left[\cot\frac{z_1}{a}\right] - HM\operatorname{Im}\left[\cot\frac{z_2}{a}\right]\right\},$$

$$\tau_{xy} = \frac{P_0 MN}{a\pi(GN - MH)}\mathrm{Re}\left[\cot\frac{z_1}{a} - \cot\frac{z_2}{a}\right].$$

When $w = 0$, all the above results become those corresponding ones in Chapter Ⅱ.

§ 2. Periodic Contact Problems with Moving Stamps

In § 4, Chapter Ⅱ, for periodic contact problems of isotropic medium, it is assumed that the periodically arranged stamps are in static equilibrium state. In the present section, we assume these stamps are moving with a constant velocity w along the boundary of the half-plane S^-. Since all the discussions are very similar to those in that section, we would only give the results without details in calculations.

1. Periodic boundary condition and solution of the problems

First, applying (1.59) and (1.60), we have, by (4.2) and (4.3),

$$\Phi(z_1) = \frac{1}{GN - MH}\frac{1}{2a\pi}\int_{L_0}[N\sigma_y(t) - iH\tau_{xy}(t)]\cot\frac{t - z_1}{a}dt, \tag{4.15}$$

$$\Psi(z_2) = \frac{-1}{GN - MH}\frac{1}{2a\pi}\int_{L_0}[M\sigma_y(t) - iG\tau_{xy}(t)]\cot\frac{t - z_2}{a}dt, \tag{4.16}$$

Taking the boundary value of

$$\frac{\partial v}{\partial x} = C\Phi(z_1) + C\overline{\Phi(z_1)} + D\Psi(z_2) + D\overline{\Psi(z_2)},$$

we have, by (4.15) and (4.16),

$$\left(\frac{\partial v}{\partial x}\right)_{y=0} = \frac{CN - DM}{GN - MH}\frac{1}{a\pi}\int_{L_0}\sigma_y(t)\cot\frac{t - x}{a}dt + \frac{DG - CH}{GN - MH}\tau_{xy}(x). \tag{4.17}$$

Next, consider the boundary condition on L_0:

$$\left.\begin{aligned} &\sigma_y = 0,\ \tau_{xy} = 0,\ x \in \gamma_0' = L_0 - \gamma_0, \\ &\frac{\partial v}{\partial x} = f'(x),\ \tau_{xy} + \rho\sigma_y = 0,\ x \in \gamma_0, \end{aligned}\right\} \tag{4.18}$$

where $y = f(x)$ is the equation of the base of the stampes, with $f'(x) \in H$ arc-wisely, and ρ is the coefficient of friction, with the given principal vector of the external stresses on γ_0: $X + iY = (\rho - i)P_0$. ρ may be positive or negative according as the velocity $w > 0$ or < 0.

Defince $w_1(z) = u_1 - iv_1$ and $w_2(z) = u_2 - iv_2$ as before, by (3.56) and (3.57) respectively. Then, (4.18) may be interpreted by $w_1(z)$ and $w_2(z)$, i.e., by (4.17).

$$v_1 = v_2 = 0, \ x \in \gamma_0',$$

$$\frac{CN - DM}{a\pi(GN - MH)} u_1 + \frac{GD - CH}{a\pi(GN - MH)} v_1 = f'(x), \ x \in \gamma_0,$$

$$v_2 + \rho v_1 = 0.$$

Thus, the periodic function $w_1(z)$ holomorphic in S^- satisfies the boundary condition on L_0:

$$a(x) u_1^-(x) + b(x) v_1^-(x) = F(x), \ x \in L_0, \tag{4.19}$$

where

$$a(x) = \begin{cases} 1, & x \in \gamma_0, \\ 0, & x \in \gamma_0'; \end{cases} \qquad b(x) = \begin{cases} \rho \dfrac{CH - DG}{CN - DM}, & x \in \gamma_0, \\ 1, & x \in \gamma_0'; \end{cases}$$

$$F(x) = \begin{cases} \dfrac{GN - MH}{CN - DM} a\pi f'(x), & x \in \gamma_0, \\ 0, & x \in \gamma_0'. \end{cases}$$

(4.19) is a Riemann-Hilbert boundary value problem of the half-plane S^- with discontinuous coefficients, of the type discussed in § 3.2, Chapter Ⅲ.

Write

$$p = \frac{GN - MH}{CN - DM}, \ q = \frac{CH - DG}{CN - DM}, \tag{4.20}$$

and

$$\theta = \frac{1}{\pi}\arctan(\rho q), \ |\theta| < \frac{1}{2}.$$

Now, as in (3.72),

$$w_1(z) = \pm ie^{\pm\theta\pi i} p\cos\theta\pi E(z) \int_{\gamma_0} \frac{f'(t)}{E(t)} \cot\frac{t - z}{a} dt$$

$$\pm ie^{-\theta\pi i}\left(C_1 \tan\frac{z}{a} + C_2\right) E(z), \ z \in S^\pm, \tag{4.21}$$

where C_1 and C_2 are undetermined constants, provided that $E(z)$ and $E(t)$ are given by (3.67) and (3.69) ($E(t) \geqslant 0$ for $t \in \gamma$) respectively. And our solution is (4.21) for $z \in S^-$.

By the boundary condition, we also have

$$w_2(z) = -\rho w_1(z) + \beta, \tag{4.22}$$

where β is an undetermined real constant.

2. Periodic condition for displacements and equilibrium condition at $z = -\infty i$

Let us find C_1, C_2 and β so as to determine the solution (4.21) ($z \in S^-$) completely, by using the periodic condition of displacements and the elastic equilibrium at $z = -\infty i$.

Similar to the previous section, $\Phi(z_1)$ and $\Psi(z_2)$ may be represented by $w_1(z)$ and $w_2(z)$:

$$\Phi(z_1) = \frac{1}{2a\pi(GN - MH)}[Nw_1(z_1) - iHw_2(z_1)],$$
$$\Psi(z_2) = \frac{1}{2a\pi(GN - MH)}[Mw_1(z_2) - iGw_2(z_2)].$$

By substitution into the periodic condition

$$[u + iv]_{\Lambda_-} = 0,$$

we get

$$(\varepsilon_1 + i\delta_1)\int_{\Lambda_-} w_1(z)dz + (\varepsilon_2 + i\delta_2)\int_{\Lambda_-} \overline{w_1(z)}dz - 2i\eta a\pi\beta = 0, \tag{4.23}$$

where we have put

$$\left.\begin{aligned}
\varepsilon_1 + i\delta_1 &= \frac{1}{GN - MH}\{(A + C)N - (B + D)M \\
&\quad + i\rho[(A + C)H - (B + D)G]\}, \\
\varepsilon_2 + i\delta_2 &= \frac{1}{GN - MH}\{(A - C)N - (B - D)M \\
&\quad + i\rho[(A - C)H - (B - D)G]\}, \\
\eta &= \frac{AH - BG}{GN - MH},
\end{aligned}\right\} \tag{4.24}$$

and Λ_- is shown in Fig.2.6.

By substituting the expression (4.21) of $w_1(z)$ into (4.23), the following equation is obtained:

$$\begin{aligned}
&[(\varepsilon_1 + i\delta_1)e^{-i\frac{2l\theta}{a}} + (\varepsilon_2 + i\delta_2)e^{i\frac{2l\theta}{a}}]C_1 \\
&+ i[(\varepsilon_1 + i\delta_1)e^{-i\frac{2l\theta}{a}} - (\varepsilon_2 + i\delta_2)e^{i\frac{2l\theta}{a}}]C_2 - \frac{2i\eta}{\cos\frac{l}{a}}\beta \\
&= -p\cos\pi\theta[(\varepsilon_1 + i\delta_1)e^{-i\frac{2l\theta}{a}} + (\varepsilon_2 + i\delta_2)e^{i\frac{2l\theta}{a}}]\int_{\gamma_0}\frac{f(t)}{E(t)}dt.
\end{aligned} \tag{4.25}$$

Consider then the condition of equilibrium at $z = -\infty i$:

$$P_0 = \mathrm{Im} w(-\infty i).$$

By suibstitution into (4.21), we get

$$-C_1 \sin\frac{2l\theta}{a} + C_2 \cos\frac{2l\theta}{a} = \frac{P_0}{\cos\frac{l}{a}} + p\cos\pi\theta\sin\frac{2l\theta}{a}\int_{\gamma_0}\frac{f'(t)}{E(t)}dt. \quad (4.26)$$

Thus, C_1, C_2 and β are uniquely determined by (4.25) and (4.26) and hence our problem is completely solved.

3. The pressure beneath the stamps

The pressure distribution beneath the stamps is

$$p(x) = -\frac{1}{a\pi}\mathrm{Im}[w_1^-(x)], \quad x \in \gamma_0. \quad (4.27)$$

Then, again by (4.21),

$$p(x) = -\frac{P}{2}\sin 2\pi\theta f'(x) + \frac{p\cos^2\pi\theta}{a\pi}E(x)\int_{\gamma_0}\frac{f'(t)}{E(t)}\cot\frac{t-x}{a}dt$$
$$+\frac{\cos\pi\theta}{a\pi}E(x)\left(C_1\tan\frac{x}{a} + C_2\right), \quad x \in \gamma_0, \quad (4.28)$$

where C_1 and C_2 have been determined before.

Example 4.5 Consider the case where the periodic bases are rectilinear. In this case, $f'(x)=0$. Then, by (4.28),

$$p(x) = \frac{\cos\pi\theta}{a\pi}E(x)\left(C_1\tan\frac{x}{a} + C_2\right), \quad (4.29)$$

where C_1 and C_2 are determined by

$$C_1\left[(\varepsilon_1+\varepsilon_2)\cos\frac{2l\theta}{a} + (\delta_1-\delta_2)\sin\frac{2l\theta}{a}\right] + C_2\left[(\delta_2-\delta_1)\cos\frac{2l\theta}{a} + (\varepsilon_1+\varepsilon_2)\sin\frac{2l\theta}{a}\right] = 0,$$
$$C_1\sin\frac{2l\theta}{a} - C_2\cos\frac{2l\theta}{a} = -\frac{P_0}{\cos\frac{l}{a}}.$$

Finally, we obtain

$$P(x) = \frac{P_0\cos\pi\theta E(x)}{a\pi(\varepsilon_1+\varepsilon_2)\cos\frac{l}{a}}\Big\{\left[(\delta_1-\delta_2)\tan\frac{x}{a} + (\varepsilon_1+\varepsilon_2)\right]\cos\frac{2l\theta}{a}$$
$$-\left[(\varepsilon_1+\varepsilon_2)\tan\frac{x}{a} - (\delta_1-\delta_2)\right]\sin\frac{2l\theta}{a}\Big\}, \quad t \in \gamma_0. \quad (4.30)$$

In case without friction, we may get the corresponding solution by putting $\rho = 0$ in (4.30) (then $\theta = 0$), i.e.,

$$p(x) = \frac{P_0}{a\pi(\varepsilon_1 + \varepsilon_2)\cos\frac{l}{a}} \frac{(\delta_1 - \delta_2)\tan\frac{x}{a} + (\varepsilon_1 + \varepsilon_2)}{(\tan\frac{l}{a} - \tan\frac{x}{a})^{\frac{1}{2}}(\tan\frac{l}{a} + \tan\frac{x}{a})^{\frac{1}{2}}}, \quad x \in \gamma_0. \tag{4.31}$$

Remark 1 In the above discussions, if $a \to \infty$, then we return to the case of a single stamp, which was studied by Galin [1].

Remark 2 The above discussions may be extended to the case where there are many stamps in a period without difficulty in principle.

Chapter V

Periodic Crack Problems in Plane Elasticity

For isotropic elastic plane weakened by a periodic row of cracks, W. T. Koiter had studied the first fundamental problems by complex variable methods for the cases where the cracks are rectilinear and collinear (in the direction of period)[1], or parallel and perpendicular to the direction of period[2], under very special assumptions for the external stresses subjected on the cracks as well as those at infinity.

In the present chapter, on the basis of § 1, Chapter II, under the assumptions that the stresses at infinity are finite and the displacements are quasi-periodic, the periodic fundamental problems are discussed in case that the periodic cracks are rectilinear and collinear. And then, the same problems are studied in the general case where the shape and number (in a period) of the cracks may be arbitrary, which are reduced to singular integral equations by using conformal mapping. At last, the similar problems in anisotropic plane elasticity for periodic collinear cracks are investigated in case of symmetric or anti-symmetric loads on the cracks.

§ 1. Fundamental Problems of Isotropic Plane with Periodic Collinear Cracks

1. General comments

Consider the isotropic elastic infinite plane weakened by periodic rectilinear cracks with the same direction as the period $a\pi$. Without loss of generality, we may ask that they are situated on the real axis.

Assume that there are n cracks in the periodic strip $|x| < \frac{1}{2}a\pi$, namely, l_k: $a_k \leqslant t \leqslant b_k (a_{k+1} > b_k)$, $k = 1, \cdots, n-1$, positively oriented from a_k to b_k. Denote $l_0 = \sum_{k=1}^{n} l_k$ and the principal vector of the external stresses on l_k by $X_k + iY_k$. The elastic region is denoted by S. Other notations are the same as in Chapter II.

The following discussions are made under the assumptions that the stresses are periodic and bounded at $z = \pm\infty i$ while the displacements are quasi-periodic.

Introduce functions

$$\left.\begin{aligned} \omega(z) &= z\bar{\Phi}(z) + \bar{\Psi}(z), \\ \Omega(z) &= \omega'(z) = \bar{\Phi}(z) + z\bar{\Phi}'(z) + \bar{\Psi}'(z), \end{aligned}\right\} \quad z \in S, \tag{5.1}$$

where $\Phi(z)$, $\Psi(z)$ are complex stress functions and $\bar{\Phi}(z) = \overline{\Phi(\bar{z})}$, $\bar{\Psi}(z) = \overline{\Psi(\bar{z})}$. It is easily seen

$$\sigma_y - i\tau_{xy} = \Phi(z) + \Omega(\bar{z}) + (z - \bar{z})\overline{\Phi'(z)}, \quad z \in S. \tag{5.2}$$

By periodicity of $\Phi(z)$, we know that $\Omega(z)$ is also periodic.

Assume both $\Phi(z)$ and $\Omega(z)$ are at most integrably unbounded at the tips of l_k and

$$\lim_{z\to t} y\Phi'(z) = 0 \qquad (z = x + iy \in S,\ t \in L_0). \tag{5.3}$$

Obviously,

$$\left.\begin{aligned} X_k &= \int_{l_k} [\tau_{xy}^-(t) - \tau_{xy}^+(t)]dt, \\ Y_k &= \int_{l_k} [\sigma_y^-(t) - \sigma_y^+(t)]dt, \end{aligned}\right\} \quad k = 1, \cdots, n, \tag{5.4}$$

where $\sigma_y^\pm + i\tau_{xy}^\pm$ are the external stresses on the upper bank and the lower bank of l_k respectively. Thereby, their resultant $X + iY$ is given by

$$\left.\begin{aligned} X &= \int_{l_0} [\tau_{xy}^-(t) - \tau_{xy}^+(t)]dt, \\ Y &= \int_{l_0} [\sigma_y^-(t) - \sigma_y^+(t)]dt. \end{aligned}\right\} \tag{5.5}$$

Let $\Phi_0(z)$ and $\Psi_0(z)$ be defined by (2.16) and (2.17)[①], Then $\Phi_0(\pm\infty i) = \Psi_0(\pm\infty i) = 0$. We may easily verify that

$$\left.\begin{aligned} \Phi(\pm\infty i) &= \mp\frac{Y - iX}{2a\pi(\kappa + 1)} + \beta, \\ \Psi(\pm\infty i) &= \mp\frac{(\kappa - 1)Y + i(\kappa + 1)X}{2a\pi(\kappa + 1)} - \beta + \kappa\bar{\beta} - q; \end{aligned}\right\} \tag{5.6}$$

and then, by (5.1),

$$\Omega(\pm\infty i) = \pm\frac{\kappa(Y - iX)}{2a\pi(\kappa + 1)} + \kappa\beta - q. \tag{5.7}$$

---

① In case of cracks, the corresponding formulas in Chapter II remain valid because we may surround the cracks in a period by a closed contour on which the principal vector of the external stresses is identical to the resultant on these cracks.

2. The first fundamental problem

Assume $\sigma_y^{\pm}(t)$ and $\tau_{xy}^{\pm}(t)$ are given, $\in H$, and σ_-, τ_-, h_- (consequently σ_+, τ_+, h_+, β, q) are also given. Find the equilibrium.

According to the boundary condition, we assure, by (5.2),

$$\Phi^+(t) + \Omega^-(t) = \sigma_y^+ - i\tau_{xy}^+,$$
$$\Phi^-(t) + \Omega^+(t) = \sigma_y^- - i\tau_{xy}^-.$$

By addition and subtraction, our problem is easily transferred to the following two boundary value problems:

$$[\Phi(t) + \Omega(t)]^+ + [\Phi(t) + \Omega(t)]^- = 2p(t), \tag{5.8}$$
$$[\Phi(t) - \Omega(t)]^+ - [\Phi(t) - \Omega(t)]^- = 2q(t), \tag{5.9}$$

where we have put

$$p(t) = \frac{1}{2}[\sigma_y^+(t) + \sigma_y^-(t)] - \frac{i}{2}[\tau_{xy}^+(t) + \tau_{xy}^-(t)], \tag{5.10}$$
$$q(t) = \frac{1}{2}[\sigma_y^+(t) - \sigma_y^-(t)] - \frac{i}{2}[\tau_{xy}^+(t) - \tau_{xy}^-(t)]. \tag{5.11}$$

We know that, by (5.5),

$$\int_{l_0} q(t)dt = -\frac{1}{2}(Y - iX). \tag{5.12}$$

By applying the generalized Plemelj formula (1.14), the solution of (5.9) in class h_0 is

$$\Phi(z) - \Omega(z) = \frac{1}{a\pi i}\int_{l_0} q(t)\cot\frac{t-z}{a}dt + 2C, \tag{5.13}$$

where C is a constant. Let $z \to \pm\infty i$ in (5.13). By addition, we get

$$C = \frac{1}{2}[q - (\kappa - 1)\beta], \tag{5.14}$$

Next, by applying the results given in § 2.1, Chapter Ⅰ, the solution of (5.8) in class h_0 is

$$\Phi(z) + \Omega(z) = \frac{X(z)}{a\pi i}\int_{l_0} \frac{p(t)}{X^+(t)}\cot\frac{t-z}{a}dt + 2X(z)P_n(\tan\frac{z}{a}), \tag{5.15}$$

where

$$P_n(\zeta) = C_0\zeta^n + \cdots + C_n \tag{5.16}$$

is a polynomial of degree n with arbitrary coefficients and

$$X(z) = \prod_{k=1}^{n}\left(\tan\frac{z}{a} - \tan\frac{a_k}{a}\right)^{-\frac{1}{2}}\left(\tan\frac{z}{a} - \tan\frac{b_k}{a}\right)^{-\frac{1}{2}}, \tag{5.17}$$

the radicals in which may be arbitrarily taken as a continuous branch in the z-plane cut by the periodic cracks, for instance, that branch fulfilling

$$\lim_{z \to \frac{a\pi}{2}} \tan^n \frac{z}{a} X(z) = 1.$$

Thus, by (5.13) and (5.15), we have

$$\left.\begin{aligned} \Phi(z) &= \frac{X(z)}{2a\pi i}\int_{l_0} \frac{p(t)}{X^+(t)}\cot\frac{t-z}{a}dt \\ &\quad + \frac{1}{2a\pi i}\int_{l_0} q(t)\cot\frac{t-z}{a}dt + X(z)P_n\left(\tan\frac{z}{a}\right) + C, \\ \Omega(z) &= \frac{X(z)}{2a\pi i}\int_{l_0} \frac{p(t)}{X^+(t)}\cot\frac{t-z}{a}dt \\ &\quad - \frac{1}{2a\pi i}\int_{l_0} q(t)\cot\frac{t-z}{a}dt + X(z)P_n\left(\tan\frac{z}{a}\right) - C. \end{aligned}\right\} \tag{5.18}$$

The rest of our problem is to determine $C_0, \cdots, C_n$.

Now, by (5.1), we have the displacement $u + iv$ expressed as

$$2\mu[u(z) + iv(z)] = \kappa\varphi(z) - \omega(\bar{z}) - (z - \bar{z})\overline{\Phi(z)} + \text{const},$$

and so, for t on the positive and the negative sides of l_0,

$$2\mu[u^{\pm}(t) + iv^{\pm}(t)] = \kappa\varphi^{\pm}(t) - [\omega^{\mp}(t)], \ t \in l_0.$$

When t describes along l_k, they give rise to

$$\begin{aligned} 2\mu[u^{\pm}(t) + iv^{\pm}(t)]_{l_k} &= \kappa[\varphi^{\pm}(t)]_{l_k} - [\omega^{\mp}(t)]_{l_k} \\ &= \kappa\int_{l_k}\Phi^{\pm}(t)dt - \int_{l_k}\Omega^{\mp}(t)dt. \end{aligned}$$

Since $u(z)$ and $v(z)$ are single-valued and have the same values at a_k and b_k respectively, we must have $[u^+(t) + iv^+(t)]_{l_k} = [u^-(t) + iv^-(t)]_{l_k}$, and so

$$\kappa\int_{l_k}[\Phi^+(t) - \Phi^-(t)]dt + \int_{l_k}[\Omega^+(t) - \Omega^-(t)]dt = 0, \ k = 1, \cdots, n, \tag{5.19}$$

which is a system of n linear equations in $C_0, \cdots, C_n$.

Moreover, when z describes the line-segments $\Lambda_{\pm}$ in Fig. 2.6 for sufficiently large $|z|$, by quasi-periodicity of the displacements, another two equations are obtained:

$$2\mu[u(z) + iv(z)]_{\Lambda_{\pm}} = \kappa[\varphi(z)]_{\Lambda_{\pm}} - [\omega(\bar{z})]_{\Lambda_{\pm}}$$

$$= \kappa \int_{\Lambda_{\pm}} \Phi(z)dz - \int_{\Lambda_{\mp}} \Omega(z)dz = a\pi q, \qquad (5.20)$$

where q is defined by (2.29). However, one of them may be derived at once from the other by the single-valued property of $u + iv$, when consider its increment (equal to zero) as z describes a cycle along the rectangle in Fig. 2.6. Thereby, (5.19) and (5.20) consist of $n+1$ independent linear equations, the unique solvability of which is evident from mechanics. Thus, our problem is solved.

As a special case, assume there is only one crack γ_0: $-l \leqslant t \leqslant l$ $(0 < l < \frac{a\pi}{2})$ in a period. In this case, (5.18) becomes

$$\left.\begin{aligned} \Phi(z) &= \frac{1}{2a\pi i\sqrt{R(z)}}\int_{-l}^{l} p(t)\sqrt{R(t)}\cot\frac{t-z}{a}dt \\ &\quad + \frac{1}{2a\pi i}\int_{-l}^{l} q(t)\cot\frac{t-z}{a}dt + \frac{C_0\tan\frac{z}{a} + C_1}{\sqrt{R(z)}} + C, \\ \Omega(z) &= \frac{1}{2a\pi i\sqrt{R(z)}}\int_{-l}^{l} p(t)\sqrt{R(t)}\cot\frac{t-z}{a}dt \\ &\quad - \frac{1}{2a\pi i}\int_{-l}^{l} q(t)\cot\frac{t-z}{a}dt + \frac{C_0\tan\frac{z}{a} + C_1}{\sqrt{R(z)}} - C, \end{aligned}\right\} \qquad (5.18)'$$

where

$$R(z) = \tan^2\frac{l}{a} - \tan^2\frac{z}{a}, \qquad (5.21)$$

with $\sqrt{R(z)}$ being taken as the branch in the z-plane cut by the periodic cracks such that, for example,

$$\lim_{z\to\pm\infty i}\sqrt{R(z)} = \pm 1/\cos\frac{l}{a}, \qquad (5.22)$$

or, what is the same, $\sqrt{R(z)}$ takes non-negative value as $z \to t \in \gamma_0$ from the upper half-plane, i.e., $\sqrt{R(t)} \geqslant 0$ in (5.18)′.

The simplest method for determining C_0 and C_1 is as follows. Substituting $z = \pm\infty i$ in the first equation of (5.18)′, adding and subtracting to each other, and then substituting (5.6), (5.12), (5.14) into the obtained results, we then get, on account of (2.19),

$$C_0 = -\frac{1}{2a\pi i}\int_{-l}^{l} p(t)\sqrt{R(t)}dt - \frac{(\tau_- + \tau_+) + i(\sigma_- + \sigma_+)}{4\cos\frac{l}{a}}, \qquad (5.23)$$

$$C_1 = \frac{\kappa - 1}{\kappa + 1} \frac{Y - iX}{4a\pi \cos \frac{l}{a}}. \tag{5.24}$$

In particular, if $X = Y = 0$ and denote $\sigma = \sigma_{\pm}$, $\tau_\infty = \tau_{\pm}$, $h_\infty = h_{\pm}$, then we have, by (2.23) and (2.22),

$$\left.\begin{aligned} \beta &= \frac{1}{4}(h_\infty + \tau_\infty) - \frac{i\tau_\infty}{\kappa + 1}, \\ q &= \frac{\kappa + 1}{4} h_\infty - \frac{3 - \kappa}{4} \sigma_\infty . \end{aligned}\right\} \tag{5.25}$$

After substitution into (5.14), (5.23) and (5.24), we get at length

$$\left.\begin{aligned} \Phi(z) &= -\frac{(\tau_\infty + i\sigma_\infty)\sin\frac{z}{a}}{2\sqrt{\sin\frac{l+z}{a}\sin\frac{l-z}{a}}} + \frac{h_\infty - \sigma_\infty}{4} + \frac{\kappa - 1}{\kappa + 1}\frac{i\tau_\infty}{2}, \\ \Omega(z) &= -\frac{(\tau_\infty + i\sigma_\infty)\sin\frac{z}{a}}{2\sqrt{\sin\frac{l+z}{a}\sin\frac{l-z}{a}}} - \frac{h_\infty - \sigma_\infty}{4} - \frac{\kappa - 1}{\kappa + 1}\frac{i\tau_\infty}{2}, \end{aligned}\right\} \tag{5.26}$$

where the radical appeared, as before, takes non-negative value as z tends to $t \in \gamma_0$ from the upper half-plane. It is easy to observe that, when the displacements are required to be periodic, i.e., $q = 0$, we must put, by (5.25),

$$h_\infty = \frac{3 - \kappa}{\kappa + 1}\sigma_\infty .$$

It may be understood as follows. When $\sigma_\infty > 0$, that is, some tension perpendicular to the direction of the period is applied at $z = \pm \infty i$, in order to keep the periodicity of the displacements, a tension with intensity $\frac{3-\kappa}{\kappa+1}$ (>0) times of that of the former in the direction of the period must be applied at $z = \pm\infty i$.

As $a \to +\infty$ in (5.26), the stress functions for the case of a single horizontal crack are obtained:

$$\left.\begin{aligned} \Phi(z) &= -\frac{\tau_\infty + i\sigma_\infty}{2\sqrt{l^2 - z^2}} z + \frac{h_\infty - \sigma_\infty}{4} + \frac{\kappa - 1}{\kappa + 1}\frac{i\tau_\infty}{2}, \\ \Omega(z) &= -\frac{\tau_\infty + i\sigma_\infty}{2\sqrt{l^2 - z^2}} z - \frac{h_\infty - \sigma_\infty}{4} - \frac{\kappa - 1}{\kappa + 1}\frac{i\tau_\infty}{2}, \end{aligned}\right\} \tag{5.26$'$}$$

where the appeared radical, as before, takes non-negative value as z tends to $t \in \gamma_0$ from the upper half-plane. The result is identical to that of Muskhelishvili.

The periodic problem for the case $n = 1$ was studied by Koiter, without external stresses on the cracks.

3. The second fundamental problem

For simplicity, we assume that there occurs only one single crack in a period, that is, γ_0: $[-l, l]$.

Given the periodic displacements $u^{\pm}(t) + iv^{\pm}(t)$ respectively on the upper and the lower banks of the cracks, with $u^{\pm\prime}(t) + iv^{\pm\prime}(t) \in H$, the resultant of the principal vectors of the external stresses on $\gamma_0 : X + iY$, and the stresses at $z = -\infty i$ (and hence at $z = +\infty i$), find the equilibrium.

It may be solved by the method analogous to that in the previous paragraph.

Taking partial derivatives with respect to x on both sides of the expression

$$2\mu(u + iv) = \kappa\varphi(z) - \omega(\bar{z}) - (z - \bar{z})\overline{\Phi(z)} + \text{const},$$

we obtain

$$2\mu(u' + iv') = \kappa\Phi(z) - \Omega(\bar{z}) - (z - \bar{z})\overline{\Phi'(z)}, \tag{5.27}$$

where $u' = \dfrac{\partial u}{\partial x}$, $v' = \dfrac{\partial v}{\partial x}$. Taking boundary values on both sides of (5.27) as $z \to t$ on γ_0 from the upper and the lower banks respectively, we have

$$\kappa\Phi^+(t) - \Omega^-(t) = 2\mu(u^{+\prime} + iv^{+\prime}),$$
$$\kappa\Phi^-(t) - \Omega^+(t) = 2\mu(u^{-\prime} + iv^{-\prime}),$$

or,

$$\left.\begin{aligned} [\kappa\Phi(t) - \Omega(t)]^+ + [\kappa\Phi(t) - \Omega(t)]^- = 2f(t), \\ [\kappa\Phi(t) + \Omega(t)]^+ - [\kappa\Phi(t) + \Omega(t)]^- = 2g(t), \end{aligned}\right\} \quad t \in \gamma_0, \tag{5.28}$$

where $f(t)$ and $g(t)$ are given functions $\in H$ on γ_0:

$$f(t) = \mu[(u^+ + u^-)' + i(v^+ + v^-)'],$$
$$g(t) = \mu[(u^+ - u^-)' + i(v^+ - v^-)'].$$

On solving them, we obtain

$$\left.\begin{aligned} \Phi(z) = {} & \frac{1}{2a\pi\kappa i\sqrt{R(z)}}\int_{-l}^{l} f(t)\sqrt{R(t)}\cot\frac{t-z}{a}dt \\ & + \frac{1}{2a\pi\kappa i}\int_{-l}^{l} g(t)\cot\frac{t-z}{a}dt + \frac{C_0\tan\frac{z}{a} + C_1}{\kappa\sqrt{R(z)}} + \beta - \frac{q}{2\kappa}, \\ \Omega(z) = {} & -\frac{1}{2a\pi i\sqrt{R(z)}}\int_{-l}^{l} f(t)\sqrt{R(t)}\cot\frac{t-z}{a}dt \\ & + \frac{1}{2a\pi i}\int_{-l}^{l} f(t)\cot\frac{t-z}{a}dt - \frac{C_0\tan\frac{z}{a} + C_1}{\sqrt{R(z)}} + \kappa\beta - \frac{q}{2}, \end{aligned}\right\} \tag{5.29}$$

where

$$\left.\begin{aligned} C_0 &= -\frac{1}{2a\pi i}\int_{-l}^{l} f(t)\sqrt{R(t)}\,dt - \frac{iq}{2\cos\frac{l}{a}}, \\ C_1 &= -\frac{\kappa(Y - iX)}{2(\kappa + 1)a\pi\cos\frac{l}{a}}. \end{aligned}\right\} \tag{5.30}$$

There was incomplete discussion by H. F. Bueckner on this problem.

Remark The corresponding mixed problem may be studied by method similar to that used here.

§ 2. Fundamental Problems of Isotropic Elastic Plane with Periodic Arbitrary Cracks

1. General discussions

Assume there are periodically arranged cracks of arbitrary shape in an infinite isotropic elastic plane.

Consider the periodic strip $|\mathrm{Re}z| < \frac{a\pi}{2}$, in which p non-intersecting curved cracks $L_k = a_k b_k$, $k = 1, \cdots, p$, lie, oriented from a_k to b_k, each of which possesses Hölder continuous curvature of order α $(> \frac{1}{2})$. Denote $L_0 = \sum_{k=1}^{p} L_k$ and L, the set of cracks periodically extented by L_0.

We assume that the stresses are periodic and bounded at infinity, and the displacements are quasi-periodic.

The principal vectors of the external stresses $X_n^{\pm}(t) + iY_n^{\pm}(t)$ on the positive and the negative banks of L_k are, respectively,

$$X_k^{\pm} + iY_k^{\pm} = \int_{L_k} [X_n^{\pm}(t) + iY_n^{\pm}(t)]ds, \quad k = 1, \cdots, p, \tag{5.31}$$

where n is the normal direction of L_k and s, the arc-length. The resultant of these two vectors on L_k is

$$X_k + iY_k = (X_k^{+} + iY_k^{+}) + (X_k^{-} + iY_k^{-}),$$

and that on the whole L_0 is

$$X + iY = \sum_{k=1}^{p} (X_k + iY_k).$$

Let us illustrate the general expressions of $\varphi(z)$ and $\psi(z)$.

In order to seperating their multi-valued parts, let

$$\zeta_k(z) = \left\{\sqrt{\frac{e^{\frac{2iz}{a}} - e^{\frac{2ia_k}{a}}}{e^{\frac{2iz}{a}} - e^{\frac{2ib_k}{a}}}} + 1\right\} \Bigg/ \left\{\sqrt{\frac{e^{\frac{2iz}{a}} - e^{\frac{2ia_k}{a}}}{e^{\frac{2iz}{a}} - e^{\frac{2ib_k}{a}}}} - 1\right\} \quad (1 \leqslant k \leqslant p). \tag{5.32}$$

This function conformally maps the strip $|\mathrm{Re}z|<\frac{1}{2}a\pi$ cut by L_k into the exterior of certain simply connected bounded region in ζ_k-plane (containing the origin of the coordinates in its interior), provided that the radical appeared in it has been chosen such that it tends to $+1$ as $z\to-\infty i$. Each simple closed contour surronding L_k in the strip is mapped to one surronding $\zeta_k=0$, preserving its orientation. Note that $\zeta_k(+\infty i)$ is finite and $\zeta_k(-\infty i)=0$. Therefore, we may take the multi-valued function

$$Z_k(z) = \log\zeta_k(z) \tag{5.33}$$

in place of $\log\sin\frac{z-z_k}{a}$ in (2.16) and (2.17), where the logarithm is taken as an arbitrary branch in the z-plane cut by a curve connecting $-\infty i$ to a_k and then running along L_k to b_k and its congruents. It is obvious that

$$Z_k'(z) = \zeta_k'(z)/\zeta_k(z) = \frac{2i}{a}e^{\frac{2iz}{a}}\Big/\sqrt{(e^{\frac{2iz}{a}}-e^{\frac{2ia_k}{a}})(e^{\frac{2iz}{a}}-e^{\frac{2ib_k}{a}})}, \tag{5.34}$$

where the radical is understood as follows: it tends to $+\infty i$ as $z\to-\infty i$, i.e.,

$$Z_k'(+\infty i) = 0,\ Z_k'(-\infty i) = \frac{2i}{a}. \tag{5.35}$$

$Z_k'(z)$ is periodic and holomorphic in the z-plane cut as above, and continuous to both sides of L_k and with singularities of order $\frac{1}{2}$ at the tips a_k, b_k.

Thus, analogous to the disscussions on stress functions in § 1, Chapter II, we have the following general expressions:

$$\varphi(z) = -\frac{1}{2\pi(\kappa+1)}\sum_{k=1}^{p}(X_k+iY_k)Z_k(z)+\varphi_0(z)+\beta z, \tag{5.36}$$

$$\begin{aligned}\psi(z) = {} & \frac{\kappa}{2\pi(\kappa+1)}\sum_{k=1}^{p}(X_k-iY_k)Z_k(z)+\frac{z}{2\pi(\kappa+1)}\sum_{k=1}^{p}(X_k+iY_k)Z_k'(z)\\ & -z\varphi_0'(z)+\psi_0(z)+(\kappa\bar{\beta}-\beta-q)z,\end{aligned} \tag{5.37}$$

where $\varphi_0(z)$ and $\psi_0(z)$ are periodic functions holomorphic in the elastic region and bounded at $z=\pm\infty i$. Then,

$$\Phi(z) = -\frac{1}{2\pi(\kappa+1)}\sum_{k=1}^{p}(X_k+iY_k)Z_k'(z)+\Phi_0(z)+\beta, \tag{5.38}$$

$$\begin{aligned}\Psi(z) = {} & \frac{\kappa}{2\pi(\kappa+1)}\sum_{k=1}^{p}(X_k-iY_k)Z_k'(z)\\ & +\frac{1}{2\pi(\kappa+1)}\sum_{k=1}^{p}(X_k+iY_k)[Z_k'(z)+Z_k''(z)]\\ & -\Phi_0(z)-z\Phi_0'(z)+\Psi_0(z)+(\kappa\bar{\beta}-\beta-q),\end{aligned} \tag{5.39}$$

where $\Phi_0(z) = \varphi_0'(z)$ and $\Psi_0(z) = \Psi_0'(z)$.

2. The stresses at infinity

First, we note that, besides (5.35), it is earily verified:

$$\lim_{z \to \pm \infty i} z Z_k''(z) = 0. \tag{5.40}$$

Secondly, now that the stresses at $z = \pm \infty i$ are bounded, we have, similar to the discussions in § 1, Chapter II,

$$\Phi_0(\pm \infty i) = \Psi_0(\pm \infty i) = 0, \tag{5.41}$$

and so, since

$$\lim_{z \to \pm \infty i} z \Phi'(z) = 0,$$

we have, by differentiating (5.38),

$$\lim_{z \to \pm \infty i} z \Phi_0'(z) = 0 \tag{5.42}$$

by virture of (5.40).

By substituting the above results into (5.38) and (5.39), it is known that

$$\left.\begin{aligned} &\Phi(+\infty i) = \beta, \ \Phi(-\infty i) = \frac{Y - iX}{a\pi(\kappa + 1)} + \beta, \\ &\Psi(+\infty i) = \kappa\bar{\beta} - \beta - q, \\ &\Psi(-\infty i) = \frac{(\kappa - 1)Y + i(\kappa + 1)X}{a\pi(\kappa + 1)} + \kappa\bar{\beta} - \beta - q. \end{aligned}\right\} \tag{5.43}$$

As before, denote the values of σ_y, τ_{xy}, σ_x at $z = \pm \infty i$ by $\sigma_\pm$, $\tau_\pm$, $h_\pm$ respectively. By (5.43) and the basic expressions of $\varphi(z)$ and $\psi(z)$, we have

$$\left.\begin{aligned} &h_+ + i\tau_+ = 2\bar{\beta} - (\kappa + 1)\beta + q, \\ &h_- + i\tau_- = \frac{(3 - \kappa)Y + i(\kappa + 1)X}{a\pi(\kappa + 1)} + 2\bar{\beta} - (\kappa + 1)\beta + q; \end{aligned}\right\} \tag{5.44}$$

$$\left.\begin{aligned} &\sigma_+ + i\tau_+ = (\kappa + 1)\bar{\beta} - q, \\ &\sigma_- + i\tau_- = \frac{Y + iX}{a\pi} + (\kappa + 1)\bar{\beta} - q, \end{aligned}\right\} \tag{5.45}$$

and hence

$$\sigma_- - \sigma_+ = \frac{Y}{a\pi}, \quad \tau_- - \tau_+ = \frac{X}{a\pi}, \tag{5.46}$$

$$h_- - h_+ = \frac{3 - \kappa}{\kappa + 1} \frac{Y}{a\pi}, \tag{5.47}$$

$$q = \frac{1}{4}[(\kappa + 1)h_+ - (3 - \kappa)\sigma_+] = \frac{1}{4}[(\kappa + 1)h_- - (3 - \kappa)\sigma_-], \tag{5.48}$$

all of which are identical to those in § 1, Chapter II.

We could not evaluate β by applying the formulas in Chapter II directly since the values $\Phi(\pm\infty i)$ are different from those there. Substituting (5.48) into (5.45), we get

$$\beta = \frac{h_+ + \sigma_+}{4} - \frac{i\tau_+}{\kappa+1} + \frac{Y - iX}{a\pi(\kappa+1)} = \frac{h_- + \sigma_-}{4} - \frac{i\tau_-}{\kappa+1} - \frac{Y - iX}{a\pi(\kappa+1)}. \tag{5.49}$$

However, all these equations may be established directly from there if we consider closed contour Γ_k surrounding L_k $(k = 1, \cdots, p)$ with L_j $(j \neq k)$ in its exterior.

3. The first fundamental problem

Assume the periodic external stresses $X_n^{\pm}(t) + iY_n^{\pm}(t)$ on the cracks are given, $\in H^*$. A function $F(t) \in H^*$ on an open arc $\gamma = ab$ means that, for $c = a$ or b,

$$F(t) = \frac{F^*(t)}{|t - c|^{\alpha}}, \quad 0 < \alpha < 1,$$

where $F^*(t) \in H$ on γ (Cf. Muskhelishvili [2], here with slight modification). Moreover, the stresses at, e.g., $z = +\infty i$ are given: σ_+, τ_+, h_+ (consequently σ_-, τ_-, h_-, β and q are known). Find the equilibrium.

Let

$$\left.\begin{aligned} f^+(t) &= f_j^+(t) = i\int_{a_j}^{t}(X_n^+ + iY_n^+)\,ds, \\ f^-(t) &= f_j^-(t) = i\int_{a_j}^{b_j}(X_n^+ + iY_n^+)\,ds \\ &\quad + i\int_{t}^{b_j}(X_n^- + iY_n^-)\,dt, \end{aligned}\right\} \quad t \in L_j,\ j = 1, \cdots, p, \tag{5.50}$$

where the integrals are taken along L_j, s being the arc-length parameter. All of them are periodically extended to L (all the cracks).

Analogous to (2.35), we want to solve the following boundary value problem:

$$\left.\begin{aligned} \varphi^+(t) + t\overline{\varphi'^+(t)} + \overline{\psi^+(t)} &= f^+(t) + C(t), \\ \varphi^-(t) + t\overline{\varphi'^-(t)} + \overline{\psi^-(t)} &= f^-(t) + C(t), \end{aligned}\right\} \quad t \in L, \tag{5.51}$$

where $C(t) = C_j$ (constant) when $t \in L_j$ (in general, $t_k \neq t_j$ when $k \neq j$). However, $C(t + a\pi) \neq C(t)$ for $t \in L$. In fact, by (5.36) and (5.37), it is easily seen,

$$\varphi(z + a\pi) - \varphi(z) = a\pi\beta,$$
$$\psi(z + a\pi) - \psi(z) = \frac{a}{2(\kappa + 1)}\sum_{j=1}^{p}(X_j + iY_j)Z_j'(z) - a\pi\varphi_0'(z) + (\kappa\bar{\beta} - \beta - q)a\pi,$$

and then, by (5.51), $C(t)$ it quasi-periodic:

$$C(t + a\pi) - C(t) = a\pi[(\kappa + 1)\beta - q] = a\pi(\sigma_+ - i\tau_+). \quad (5.52)$$

Denote $C(t) = C_0(t)$ when $t \in L_0$.

By substituting (5.36) and (5.37) into (5.51), the boundary value problem (5.51) is transferred to the following periodic boundary value problem for holomorphic functions $\varphi_0(z)$ and $\psi_0(z)$:

$$\varphi_0^{\pm}(t) + (t - \bar{t})\,\overline{\varphi_0'^{\pm}(t)} + \overline{\psi_0^{\pm}(t)} = f_0^{\pm}(t) + \nu(t), \quad t \in L, \quad (5.53)$$

where we have set

$$f_0^{\pm}(z) = f^{\pm}(z) + \frac{1}{2\pi(\kappa + 1)}\sum_{j=1}^{p}(X_j + iY_j)[Z_j^{\pm}(t) - \kappa\overline{Z_j^{\pm'}(t)}] + \frac{t - \bar{t}}{2\pi(\kappa + 1)}\sum_{j=1}^{p}(X_j - iY_j)\,\overline{Z_j^{\pm'}(t)}, \quad (5.54)$$

$$\nu(t) = C(t) - (\beta + \bar{\beta})t - (\kappa\beta - \bar{\beta} - q)\bar{t} = C(t) - \frac{h_+ + \sigma_+}{2}t + (\frac{h_+ - \sigma_+}{2} + i\tau_+)\bar{t}, \quad (5.55)$$

which is already $a\pi$-periodic by (5.52).

We always assume

$$\varphi_0(-\infty i) = \psi_0(-\infty i) = 0 \quad (5.56)$$

and $\varphi_0(+\infty i)$, $\psi_0(+\infty i)$ are finite.

We also assume, for the time being, $\varphi_0^{\pm}(t) \in H$, $\varphi_0^{\pm'}(t)$ and $\psi_0^{\pm'}(t) \in H^*$. These assumptions will be verified to be true later.

By the conformal mapping

$$w = e^{\frac{2iz}{a}}, \text{ i.e., } z = \frac{a}{2i}\log w \quad (5.57)$$

(write $\tau = e^{\frac{2it}{a}}$ for $t \in L_0$ or $t = \frac{a}{2i}\log\tau$, $|\arg\tau| < \pi$), the elastic plane is mapped to the w-plane, with $z = +\infty i$ and $z = -\infty i$ to $w = 0$ and $w = \infty$ respectively. Denote

$$\alpha_j = e^{\frac{2ia_j}{a}}, \quad \beta_j = e^{\frac{2ib_j}{a}} \quad (j = 1, \cdots, p),$$

$\Gamma_j = \alpha_j\beta_j$ as the image of L_j and $\Gamma = \sum_{j=1}^{p}\Gamma_j$. Then, $\varphi_0(z)$ and $\psi_0(z)$ become sectionally holomorphic functions in the w-plane:

$$\varphi_0^*(w) = \varphi_0\left(\frac{a}{2i}\log w\right), \quad \psi_0^*(w) = \psi_0\left(\frac{a}{2i}\log w\right) \tag{5.58}$$

respectively, with jumping curve Γ. Moreover,

$$\varphi_0^{\pm\prime}(t) = \varphi_0^{*\pm\prime}(\tau)\frac{d\tau}{dt} = \frac{2i}{a}\tau\varphi_0^{*\pm\prime}(\tau).$$

Denote $C_0^*(\tau) = C_0(t)$, i.e., $C^*(\tau) = C_j$ when $\tau \in \Gamma_j$. Then, (5.53) becomes

$$\varphi_0^\pm(\tau) - 2\bar{\tau}\ln|\tau|\overline{\varphi_0^{\pm\prime}(\tau)} + \overline{\psi_0^\pm(\tau)} = f_0^{*\pm}(\tau) - \frac{a}{2}\{(\tau_+ - ih_+)\ln|\tau|$$
$$+ (\sigma_+ - i\tau_+)\arg\tau\} + C_0^*(\tau), \quad \tau \in \Gamma \quad (|\arg\tau| < \pi), \tag{5.59}$$

on account of (5.55), where

$$f_0^{*\pm}(\tau) = f_0^\pm\left(\frac{2i}{a}\log\tau\right) = f^\pm\left(\frac{2i}{a}\log\tau\right)$$
$$+ \frac{1}{2\pi(\kappa+1)}\sum_{k=1}^{p}(X_k + iY_k)[Z_k^{*\pm}(\tau) - \kappa\overline{Z_k^{*\pm}(\tau)}]$$
$$- \frac{\bar{\tau}\ln|\tau|}{\pi(\kappa+1)}\sum_{k=1}^{p}(X_k - iY_k)\overline{Z_k^{*\pm\prime}(\tau)}, \tag{5.60}$$

in which

$$Z_k^*(w) = \log\zeta_k^*(w),$$
$$\zeta_k^*(w) = \left\{\sqrt{\frac{w-\alpha_k}{w-\beta_k}} + 1\right\}\Big/\left\{\sqrt{\frac{w-\alpha_k}{w-\beta_k}} - 1\right\},$$

the single-valued branches of the involved radical and the logarithm being uniquely chosen before (the value of the radical is $+1$ at $w = \infty$). In the mean while, (5.56) becomes

$$\varphi_0^*(\infty) = \psi_0^*(\infty) = 0. \tag{5.61}$$

Moreover, we note that $\varphi_0^*(w)$ and $\psi_0^*(w)$ remain bounded at $w = 0$ since Γ does not pass by $w = 0$.

Thus, our problem is reduced to the boundary value problem (5.59) in the w-plane, reguired to satisfy the supplementary condition (5.61), in which, $\varphi_0^* \in H$ and $\varphi_0^{*\prime}$, $\psi_0^{*\prime} \in H^*$ by assumption.

The boundary condition (5.59) may be rewritten as

$$\left.\begin{aligned} &[\overline{\varphi_0^{*+}(\tau)} - \overline{\varphi_0^{*-}(\tau)}] - 2\tau\ln|\tau|[\varphi_0^{*+\prime}(\tau) - \varphi_0^{*-\prime}(\tau)] \\ &\quad + [\psi_0^{*+}(\tau) - \psi_0^{*-}(\tau)] = F_0^*(\tau), \\ &[\overline{\varphi_0^{*+}(\tau)} - \overline{\varphi_0^{*-}(\tau)}] - 2\tau\ln|\tau|[\varphi_0^{*+\prime}(\tau) + \varphi_0^{*-\prime}(\tau)] \\ &\quad + [\psi_0^{*+}(\tau) + \psi_0^{*-}(\tau)] = G_0^*(\tau) - a\{(\tau_+ + ih_+)\ln|\tau| \\ &\quad + (\sigma_+ + i\tau_+)\arg\tau\} + 2\overline{C_0^*(\tau)}, \end{aligned}\right\} \quad \tau \in \Gamma, \tag{5.62}$$

where, for $\tau \in \Gamma_j$,

$$\left.\begin{aligned} F_0^*(\tau) &= \overline{F_j^*(\tau)} - \frac{2\tau \ln|\tau|}{\pi(\kappa+1)} \frac{X_j + iY_j}{\sqrt{(\tau-\alpha_j)(\tau-\beta_j)}}, \quad w \\ G_0^*(\tau) &= \overline{f_j^{*+}(\tau)} + \overline{f_j^{*-}(\tau)} + \frac{1}{\pi(\kappa+1)} \sum_{k \neq j} (X_k - iY_k)[\overline{Z_k^{*+}(\tau)} \\ &\quad - \kappa \overline{Z_k^{*-}(\tau)}] - \frac{2\tau \ln|\tau|}{\pi(\kappa+1)} \sum_{k \neq j} \frac{X_k + iY_k}{\sqrt{(\tau-\alpha_k)(\tau-\beta_k)}}, \end{aligned}\right\} \tag{5.63}$$

in which we have put

$$\begin{aligned} F^*(\tau) = F_j^*(\tau) &= f_j^{*+}(\tau) - f_j^{*-}(\tau) \\ &+ \frac{X_j + iY_j}{\pi(\kappa+1)}[Z_j^{*+}(\tau) - \kappa \overline{Z_j^{*-}(\tau)}], \quad \tau \in \Gamma_j. \end{aligned}$$

The radical in (5.63) has to be understood by the limit of

$$\sqrt{(w-\alpha_k)(w-\beta_k)} = (w-\beta_k)\sqrt{\frac{w-\alpha_k}{w-\beta_k}}$$

as w tends to the point $t \in \Gamma_k$ from its positive side.

By the assumptions on $X_k^{\pm}(t) + iY_k^{\pm}(t)$, it is easily seen that $f_j^{*\pm}(\tau) \in H$, $f_j^{*\pm'}(\tau) \in H^*$ and, by (5.50),

$$\lim_{\tau \to \alpha_j} F_j^*(\tau) = 0, \ \lim_{\tau \to \beta_j} F_j^*(\tau) = 0, \ j = 1, \cdots, p.$$

Furthermore, by applying

$$\begin{aligned} \int_\Gamma \frac{\tau \ln|\tau| \, d\tau}{\sqrt{(\tau-\alpha_j)(\tau-\beta_j)}(\tau-\tau_0)} &= \int_\Gamma \frac{\tau \ln|\tau| - \tau_0 \ln|\tau_0|}{\sqrt{(\tau-\alpha_j)(\tau-\beta_j)}(\tau-\tau_0)} d\tau \\ &+ \tau_0 \ln|\tau_0| \int_\Gamma \frac{d\tau}{\sqrt{(\tau-\alpha_j)(\tau-\beta_j)}(\tau-\tau_0)}, \end{aligned}$$

it is easily verified that the above integral $\in H$ and its derivative $\in H^*$ (Cf. Muskhelishvili [2]). Therefore, if a new unknown function $\rho(\tau) \in H$ is introduced by

$$\varphi_0^*(w) = \frac{1}{2\pi i} \int_\Gamma \frac{\overline{\rho(\tau)}}{\tau - w} d\tau, \tag{5.64}$$

we may prove that $\rho'(\tau) \in H^*$ and $\rho(\alpha_j) = \rho(\beta_j) = 0$, $j = 1, \cdots, p$. Then, the boundary value problem (5.62) is transformed to the following singular integral equation

$$\frac{1}{\pi i} \int_\Gamma \frac{\rho(\tau)}{\tau - \tau_0} d\tau + \frac{1}{2\pi i} \int_\Gamma \rho(\tau) d\log \frac{\bar{\tau} - \bar{\tau}_0}{\tau - \tau_0} + \frac{1}{\pi i} \int_\Gamma \overline{\rho(\tau)} d \frac{\tau \ln|\tau| - \tau_0 \ln|\tau_0|}{\tau - \tau_0}$$

$$
= \frac{1}{2\pi i}\int_{\Gamma}\frac{F_0^*(\tau)}{\tau-\tau_0}d\tau - \frac{1}{2}G_0^*(\tau_0) + \frac{a}{2}\{(\tau_+ + ih_+)\ln|\tau_0| + (\sigma_+ + i\tau_+)\arg\tau_0\}
$$
$$
- C_0^*(\tau_0),\quad \tau_0 \in \Gamma, \tag{5.65}
$$

the right-hand member of which $\in H$, and its derivative, $\in H^*$. Or, it may be rewritten as

$$
\begin{aligned}
\boldsymbol{K}_1\rho \equiv{}& \frac{1}{\pi i}\int_{\Gamma}\frac{\rho(\tau)}{\tau-\tau_0}d\tau + \frac{1}{2\pi i}\int_{\Gamma}\rho(\tau)d\log\frac{\bar{\tau}-\bar{\tau}_0}{\tau-\tau_0} \\
&+ \frac{\tau_0}{\pi i}\int_{\Gamma}\overline{\rho(t)}\,d\,\frac{\ln|\tau|-\ln|\tau_0|}{\tau-\tau_0} \\
={}& \frac{a}{2\pi i}\int_{\Gamma}\frac{F_0(\tau)}{\tau-\tau_0}d\tau - \frac{1}{2}G_0^*(\tau_0) + \frac{a}{2}\{(\tau_+ + ih_+)\ln|\tau_0| \\
&+ (\sigma_+ + i\tau_+)\arg\tau_0\} - C_0^*(\tau_0),\quad \tau_0 \in \Gamma.
\end{aligned} \tag{5.66}
$$

We see that, the integral appeared in the second term on the left side of (5.66) is an integral with Fredholm kernel. We would explain that so does that in the third term. Since

$$
d\,\frac{\log\tau - \log\tau_0}{\tau-\tau_0} = -\frac{\log\tau - \log\tau_0 - \dfrac{1}{\tau}(\tau-\tau_0)}{(\tau-\tau_0)^2}d\tau,
$$

the coefficient of $d\tau$ on the right side of which tends to $\dfrac{1}{2\tau_0^2}$ as $\tau \to \tau_0$, and

$$
d\,\frac{\log\bar{\tau} - \log\bar{\tau}_0}{\tau-\tau_0} = \frac{\bar{\tau}-\bar{\tau}_0}{\tau-\tau_0}d\,\overline{\left(\frac{\log\tau-\log\tau_0}{\tau-\tau_0}\right)} + \overline{\left(\frac{\log\tau-\log\tau_0}{\tau-\tau_0}\right)}d\,\frac{\bar{\tau}-\bar{\tau}_0}{\tau-\tau_0}.
$$

By adding these two eqalities, our conclusion is established.

Therefore, (5.66) is a singular integral equation of normal type, required to be solved in class h_{2p}, i.e., solution of which must be bounded at all the tips of Γ (for notation, Cf. Muskhelishvili [2]). We mention that, by applying method similar to that of Muskhelishvili, it is not hard to prove the unique existence of the solution of the first fundamental problem, as well as the second one explored in the next paragraph. Moreover, since Γ possesses propeties similar to L_0, it is also easy to prove that (5.66) has a unique solution $\rho(\tau)$ after the constant $C_0^*(\tau_0) = C_j^*$, $\tau_0 \in \Gamma_j$, $j = 1, \cdots, p$, are suitably (and uniquely) chosen. $\varphi_0^*(w)$ may be found finally through (5.64), while $\psi_0(w)$ may be evaluated by the expression:

$$
\psi_0^*(w) = \frac{1}{2\pi i}\int_{\Gamma}[\rho(\tau) - 2\tau\ln|\tau|\,\overline{\rho(\tau)'}]\frac{d\tau}{\tau-w} + \frac{1}{2\pi i}\int_{\Gamma}\frac{F_0^*(w)}{\tau-w}d\tau. \tag{5.67}
$$

When return to the z-plane, the problem is completely solved.

4. The second fundamental problem

Now assume the quasi-periodic displacements $u^{\pm}(t) + iv^{\pm}(t) \in H$, with their derivatives $\in H^*$, are given (then the addend q in (5.31) is known) and also σ_+, τ_+ as well as $X + iY$ are given. Then, by (5.44) – (5.48), β, σ_-, τ_-, $h_\pm$ are known as well (or, what is the same, σ_+, τ_+, h_+, $X^+ + iY^+$ are given in advance). Certainly, we should assume

$$\left.\begin{array}{l} u^+(a_j) = u^-(a_j),\ v^+(a_j) = v^-(a_j), \\ u^+(b_j) = u^-(b_j),\ v^+(b_j) = v^-(b_j). \end{array}\right\} \quad j = 1, \cdots, p.$$

Find the equilibrium.

In this case, the problem is transformed to solve the boundary value problem

$$-\kappa\varphi^{\pm}(t) + t\overline{\varphi'^{\pm}(t)} + \overline{\psi^{\pm}(t)} = f^{\pm}(t),\ t \in L, \tag{5.68}$$

in class h_{2p}, where

$$f^{\pm}(t) = -2\mu[u^{\pm}(t) + iv^{\pm}(t)]. \tag{5.69}$$

We remind that functions on both sides of (5.68) are quasi-periodic and get the same increment $-a\pi q$ when t changes to $t + a\pi$.

We may always assume $\varphi_0(-\infty i) = 0$. However, we could not make any restriction on $\psi_0(-\infty i)$ beforehand except its boundedness, since any rigid motion is not permitted.

By substituting the general expressions (5.39) and (5.37) into (5.68), the following boundary condition is obtained:

$$-\kappa\varphi_0^{\pm}(t) + (t - \bar{t})\overline{\varphi_0^{\pm\prime}(t)} + \overline{\psi_0^{\pm}(t)} = f_0^{\pm}(t),\ t \in L, \tag{5.70}$$

where

$$\begin{aligned} f_0^{\pm}(t) = {} & f^{\pm}(t) + q\bar{t} - \frac{\kappa}{\pi(\kappa+1)}\sum_{k=1}^{p}(X_k + iY_k)\ln|\zeta_k^{\pm}(t)| \\ & + \frac{t - \bar{t}}{2\pi(\kappa+1)}\sum_{k=1}^{p}(X_k - iY_k)\overline{Z_k'^{\pm}(t)} \\ & + (\kappa\beta - \bar{\beta})(t - \bar{t}). \end{aligned} \tag{5.71}$$

Note that, at this time,

$$f_1^{\pm}(t) = f^{\pm}(t) + q\bar{t} \tag{5.72}$$

are periodic and so do $f_0^{\pm}(t)$. That means (5.70) is a periodic boundary value problem.

Under the conformal mapping (5.57), (5.71) is changed to

$$-\kappa\varphi_0^{*\pm}(\tau) - 2\bar{\tau}\ln|\tau|\overline{\varphi_0^{*\prime\pm}(\tau)} + \overline{\psi_0^{*\pm}(\tau)} = f_0^{*\pm}(\tau),\ \tau \in \Gamma, \tag{5.73}$$

where, by (5.71),

$$
\begin{aligned}
f_0^{*\pm}(\tau) = {} & f_1^{*\pm}(\tau) - \frac{\kappa}{\pi(\kappa+1)}\sum_{k=1}^{p}(X_k + iY_k)\,|\,\zeta_k^{*\pm}(\tau)\,| \\
& - \frac{\bar{\tau}\ln|\tau|}{\pi(\kappa+1)}\sum_{k=1}^{p}\overline{(X_k - iY_k)\,Z_k^{*'\pm}(\tau)} \\
& + ai(\kappa\beta - \bar{\beta})\ln|\tau|,
\end{aligned} \tag{5.74}
$$

in which we have set

$$
f_1^{*\pm}(\tau) = f^{*}\left(\frac{a}{2i}\log\tau\right) - \frac{aq}{\pi i}\,\overline{\log\tau},
$$

where the logarithm is taken as an arbitrary branch, e.g., for $\operatorname{Im}\log\tau = \arg\tau$, $|\arg\tau| < \pi$.

Thus, our problem is reduced to : solve the boundary value problem (5.73) in class h_{2p} with the supplementary condition $\varphi_0^*(\infty) = 0$. Here, again assume $\varphi_0^* \in H$ and $\varphi_0^{'*}$, $\psi_0^* \in H^*$. At this time, X_1, Y_1, $\cdots$, X_{p-1}, Y_{p-1} are undetermined constants (X_p and Y_p may be determined by these constants as $X + iY$ is given).

For clarity, rewrite (5.73) as

$$
\left.
\begin{aligned}
& -\kappa[\overline{\varphi_0^{*+}(\tau)} - \overline{\varphi_0^{*-}(\tau)}] - 2\tau\ln|\tau|\,[\varphi_0^{*'+}(\tau) - \varphi_0^{*'-}(\tau)] \\
& \quad + [\psi_0^{*+}(\tau) - \psi_0^{*-}(\tau)] = F_0^*(\tau), \\
& -\kappa[\overline{\varphi_0^{*+}(\tau)} + \overline{\varphi_0^{*-}(\tau)}] - 2\tau\ln|\tau|\,[\varphi_0^{*'+}(\tau) + \varphi_0^{*'-}(\tau)] \\
& \quad + [\psi_0^{*+}(\tau) + \psi_0^{*-}(\tau)] = G_0^*(\tau).
\end{aligned}
\right\} \tag{5.75}
$$

On account of (5.74), when $\tau \in \Gamma_j$,

$$
\begin{aligned}
F_0^*(\tau) = {} & \overline{F^*(\tau)} - \frac{2\tau\ln|\tau|}{\pi(\kappa+1)}\,\frac{X_j + iY_j}{\sqrt{(\tau-\alpha_j)(\tau-\beta_j)}}, \\
G_0^*(\tau) = {} & \overline{f_1^{*+}(\tau)} + \overline{f_1^{*-}(\tau)} - \frac{2\kappa}{\pi(\kappa+1)}\sum_{k\neq j}(X_k - iY_k)\ln|\,\zeta_k^{*+}(\tau)\,| \\
& - \frac{2\tau\ln|\tau|}{\pi(\kappa+1)}\sum_{k\neq j}\frac{X_k + iY_k}{\sqrt{(\tau-\alpha_k)(\tau-\beta_k)}} + 2ai(\kappa\beta - \bar{\beta})\ln|\tau|,
\end{aligned} \tag{5.76}
$$

where we have set

$$
\begin{aligned}
F^*(\tau) = {} & \overline{f_1^{*+}(\tau)} - \overline{f_1^{*-}(\tau)} - \frac{2\kappa}{\pi(\kappa+1)}\sum_{k=1}^{p}(X_k + iY_k)\ln|\,\zeta_k^{*+}(\tau)\,| \\
= {} & \overline{f^{+}\left(\frac{a}{2i}\log\tau\right)} - \overline{f^{-}\left(\frac{a}{2i}\log\tau\right)} \\
& - \frac{2\kappa}{\pi(\kappa+1)}\sum_{k=1}^{p}(X_k + iY_k)\ln|\,\zeta_k^{*+}(\tau)\,|, \quad \tau \in \Gamma.
\end{aligned} \tag{5.77}
$$

As in the previous paragraph, let $\varphi_0^*(w)$ be expressed as (5.64). Similardy, by (5.75), we obtain the singular integral equation

$$\frac{1}{\pi i}\int_{\Gamma}\frac{\rho(\tau)}{\tau-\tau_0}d\tau+\frac{1}{2\pi i}\int_{\Gamma}\rho(\tau)d\log\frac{\bar{\tau}-\bar{\tau}_0}{\tau-\tau_0}-\frac{1}{\kappa\pi i}\int_{\Gamma}\overline{\rho(\tau)}d\frac{\tau\ln|\tau|-\tau_0\ln|\tau_0|}{\tau-\tau_0}$$
$$=-\frac{1}{2\kappa\pi i}\int_{\Gamma}\frac{F_0^*(\tau)d\tau}{\tau-\tau_0}-\frac{1}{2\kappa}G_0^*(\tau)-\frac{1}{\kappa}(D_1+iD_2),$$

where D_1+iD_2 is an undetermined complex constant, which may be written in the following form:

$$\mathbf{K}_2\rho\equiv\frac{1}{\pi i}\int_{\Gamma}\frac{\rho(\tau)}{\tau-\tau_0}d\tau+\frac{1}{2\pi i}\int_{\Gamma}\rho(\tau)d\log\frac{\bar{\tau}-\bar{\tau}_0}{\tau-\tau_0}$$
$$-\frac{\tau_0}{\kappa\pi i}\int_{\Gamma}\overline{\rho(\tau)}d\frac{\ln|\tau|-\ln|\tau_0|}{\tau-\tau_0}$$
$$=-\frac{1}{2\kappa\pi i}\int_{\Gamma}\frac{F_0^*(\tau)d\tau}{\tau-\tau_0}+\frac{1}{2\kappa}G_0^*(\tau)+\frac{1}{\kappa}(D_1^*+iD_2^*),\quad \tau\in\Gamma, \tag{5.78}$$

where

$$D_1^*+iD_2^*=\frac{1}{\pi i}\int_{\Gamma}\overline{\rho(\tau)}d\ln|\tau|-(D_1+iD_2). \tag{5.79}$$

Obviously, (5.78) is a singular integral equation of normal type, containing $2p$ undetermined real constants $X_1, Y_1, \cdots, X_{p-1}, Y_{p-1}, D_1^*$ and D_2^*. Similar to the proof in the previous paragraph, we may establish its unique solvability in class h_{2p}.

After $\rho(\tau)$ is obtained, $\varphi_0^*(w)$ will be given by (5.64), while $\psi_0^*(w)$ may be evaluated by:

$$\psi_0^*(w)=-\frac{1}{2\pi i}\int_{\Gamma}[\kappa\rho(\tau)+2\tau\ln|\tau|\overline{\rho(\tau)'}]\frac{d\tau}{\tau-w}$$
$$+\frac{1}{2\pi i}\int_{\Gamma}\frac{F_0^*(\tau)}{\tau-w}d\tau+D_1+iD_2,$$

where D_1+iD_2 is the constant determined by (5.79). By returning to the z-plane, the problem is solved at length.

If the displacements on both sides of each crack in a period are given up to a constant term (different for different cracks) but supplementarily given the periodic resultant of the principal vectors of the external stresses on both sides of each crack, the equation (5.78) would be more similar in its form to that in the previous paragraph.

Example 5.1 Consider the case of a periodic set of horizontal collinear rectilinear cracks.

Assume $L=\sum_{k=1}^{p}L_k$ is a set of cracks on the real axis in the strip $|\mathrm{Re}z|<\frac{1}{2}a\pi$.

Here $|\tau|=1$ so that

$$\frac{\bar{\tau} - \bar{\tau}_0}{\tau - \tau_0} = - \frac{1}{\tau\tau_0},$$

and then (5.66) becomes

$$\frac{1}{\pi i}\int_{\Gamma} \frac{\rho(\tau)}{\tau - \tau_0} d\tau = \frac{1}{2\pi i}\int_{\Gamma} \frac{F_0^*(\tau)}{\tau - \tau_0} d\tau - \frac{1}{2} G_0^*(\tau_0)$$
$$+ \frac{a}{2}(\sigma_+ + i\tau_+)\arg\tau_0 + \frac{a}{2}(\tau_+ - i\sigma_+)\log\tau_0$$
$$- C_2^*(\tau_0), \quad |\arg\tau_0| < \pi, \tag{5.80}$$

where $C_2^*(\tau_0) = C_j''$, $\tau_0 \in \Gamma_j$, $j = 1, \cdots, p$, are new undetermined constants.

This is a modified convolution problem for Cauchy-type integrals, which may be solved by the method afforded by Muskhelishvili or by transferring it directly to a simple Riemann boundary value prlblem, and finally the solution may be obtained in closed integral form. When $p = 1$, it is identical to (5.18)′. We would not go to details.

Example 5.2 The case of periodic set of vertical collinear rectilinear cracks.

Assume now $L = \sum_{k=1}^{p} L_k$ is a set of cracks on the imaginary axis in the strip $|\mathrm{Re} z| < \frac{1}{2} a\pi$.

In this case, (5.66) becomes

$$\frac{1}{\pi i}\int_{\Gamma} \frac{\rho(\tau)}{\tau - \tau_0} d\tau + \frac{\tau_0}{\pi i}\int_{\Gamma} \overline{\rho(\tau)} d \frac{\ln\tau - \ln\tau_0}{\tau - \tau_0}$$
$$= \frac{1}{2\pi i}\int_{\Gamma} \frac{F_0^*(\tau)}{\tau - \tau_0} d\tau - \frac{1}{2} G_0^*(\tau_0)$$
$$+ \frac{a}{2}(\tau_+ + ih_+)\ln\tau_0 - C_1^*(\tau_0), \quad \tau_0 \in \Gamma. \tag{5.81}$$

To solve this equation, one may split it into two equations satisfied by the real and the imaginary parts of $\rho(\tau)$ and then solved by approximation methods (Cf. e.g., Baker [1]).

§ 3. Fundamental Problems of Anisotropic Elastic Plane with Periodic Collinear Cracks

1. General comments

Assume that, in the anisotropic infinite elastic plane, there are periodically arranged rectilinear cracks L_j, $j = 0, \pm 1, \pm 2, \cdots$ lying on the x-axis, each of which has the length $2l\,(l < \frac{1}{2} a\pi)$, L_0 being the interval $[-l, l]$, as shown in Fig. 5.1. Denote $L = \sum_{j=-\infty}^{\infty} L_j$ and its complement L'.

Assume that there exist periodic external loads on both sides of the cracks but no external stresses at infinity. We would study respectively the cases where the loads are symmetric or anti-symmetric on L_0.

The following discussions are made under the basic assumptions that both the stresses and displacements are periodic and the stresses at infinity are bounded.

Under these basic assumptions, by Lemma 3.1, both the stress functions $\Phi(z_1)$ and
$\Psi(z_2)$ are periodic (for all the notations, see Chapter Ⅲ).

At $z = \pm\infty i$, the principal vectors $X(\pm\infty i) + iY(\pm\infty i)$ of external stresses, by assumption, are zeros:

$$X(\pm\infty i) + iY(\pm\infty i) = a\pi[\tau_{xy}(\pm\infty i) + i\sigma_y(\pm\infty i)] = 0. \tag{5.82}$$

Our discussions may be restricted in the periodic strip $|\mathrm{Re}z| < \frac{1}{2}a\pi$.

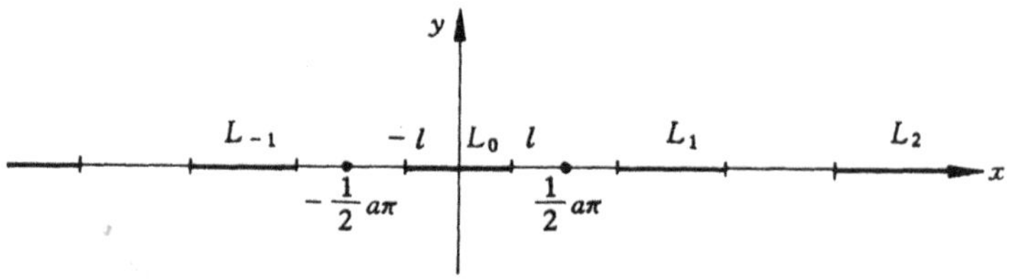

Fig.5.1

2. Symmetric loads

Assume $\Phi_1(z_1)$ and $\Psi_1(z_2)$ are the stress functions corresponding to the case where the periodic loads (in the plane) applied on the cracks are symmetric to the x-axis, i.e., tension loads only.

On the real axis, $z_1 = z_2 = \tau$. On account of symmetry, by (3.4),

$$\mu_1\Phi_1(\tau) + \mu_2\Psi_1(\tau) = 0, \quad \tau \in L'. \tag{5.83}$$

Assume $\sigma_y^{\pm}(\tau)$, $\tau \in L$, are given (periodic). By (3.3) and (5.83), the boundary condition may be expressed in $\Phi_1(z)$ only:

$$\left.\begin{aligned} \sigma_y^+(\tau) &= \frac{\mu_2 - \mu_1}{\mu_2}\Phi_1^+(\tau) + \frac{\bar{\mu}_2 - \bar{\mu}_1}{\bar{\mu}_2}\bar{\Phi}_1^-(\tau), \\ \sigma_y^-(\tau) &= \frac{\mu_2 - \mu_1}{\mu_2}\Phi_1^-(\tau) + \frac{\bar{\mu}_2 - \bar{\mu}_1}{\bar{\mu}_2}\bar{\Phi}_1^+(\tau), \end{aligned}\right\} \quad \tau \in L_0, \tag{5.84}$$

or,

$$\left[\frac{\mu_2-\mu_1}{\mu_2}\Phi_1(\tau)+\frac{\bar{\mu}_2-\bar{\mu}_1}{\bar{\mu}_2}\bar{\Phi}_1(\tau)\right]^+ + \left[\frac{\mu_2-\mu_1}{\mu_2}\Phi_1(\tau)+\frac{\bar{\mu}_2-\bar{\mu}_1}{\bar{\mu}_2}\bar{\Phi}_1(\tau)\right]^- = 2f_1(\tau),\quad \tau\in L_0, \tag{5.85}$$

$$\left[\frac{\mu_2-\mu_1}{\mu_2}\Phi_1(\tau)-\frac{\bar{\mu}_2-\bar{\mu}_1}{\bar{\mu}_2}\bar{\Phi}_1(\tau)\right]^+ - \left[\frac{\mu_2-\mu_1}{\mu_2}\Phi_1(\tau)-\frac{\bar{\mu}_2-\bar{\mu}_1}{\bar{\mu}_2}\bar{\Phi}_1(\tau)\right]^- = 2g_1(\tau),\quad \tau\in L_0, \tag{5.86}$$

where

$$f_1(\tau)=\frac{1}{2}(\sigma_y^+ + \sigma_y^-),\quad g_1(\tau)=\frac{1}{2}(\sigma_y^+ - \sigma_y^-),\quad \tau\in L.$$

Assume both $f_1(\tau)$ and $g_1(\tau)\in H$. Evidently, they are periodic. $\Phi_1(\pm\infty i)$ and $\Psi_1(\pm\infty i)$ are finite by the assumption that the stresses at infinity are bounded.

By the generalized Plemelj formula (1.14), the general solution of the boundary value problem (5.86) in class h_0 is

$$\frac{\mu_2-\mu_1}{\mu_2}\Phi_1(z_1)-\frac{\bar{\mu}_2-\bar{\mu}_1}{\bar{\mu}_2}\bar{\Phi}_1(z_1)=\frac{1}{a\pi i}\int_{-l}^{l}g_1(\tau)\cot\frac{\tau-z_1}{a}d\tau+2\beta, \tag{5.87}$$

where β is an arbitrary complex constant.

Introduce the canonical function

$$X(z)=\sqrt{R(z)},\quad R(z)=\tan^2\frac{l}{a}-\tan^2\frac{z}{a},$$

with the branch in the z-plane cut by L chosen such that

$$\lim_{z\to\pm\frac{a\pi}{2}}\frac{\tan\frac{z}{a}}{i\sqrt{R(z)}}=1.$$

By (1.27), the general solution of the periodic Riemann boundary value problem (5.85), bounded at $z=\pm\infty i$, is

$$\frac{\mu_2-\mu_1}{\mu_2}\Phi_1(z_1)+\frac{\bar{\mu}_2-\bar{\mu}_1}{\bar{\mu}_2}\bar{\Phi}_1(z_1)$$

$$=\frac{1}{a\pi i\sqrt{R(z_1)}}\int_{-l}^{l}f_1(\tau)\sqrt{R(\tau)}\cot\frac{\tau-z_1}{a}d\tau+\frac{2c_0\tan\frac{z_1}{a}+2c_1}{\sqrt{R(z_1)}}, \tag{5.88}$$

where c_0 and c_1 are arbitrary real constants.

By (5.87) and (5.88), we readily have

$$\Phi_1(z_1)=\Phi_1^*(z_1)+\frac{\mu_2}{\mu_2-\mu_1}\frac{c_0\tan\frac{z_1}{a}+c_1}{\sqrt{R(z_1)}}+\frac{\mu_2\beta}{\mu_2-\mu_1}, \tag{5.89}$$

where

$$\frac{\mu_2 - \mu_1}{\mu_2}\Phi_1^*(z_1) = \frac{1}{2a\pi i}\int_{-l}^{l} g_1(\tau)\cot\frac{\tau - z_1}{a}d\tau + \frac{1}{2a\pi i\sqrt{R(z_1)}}\int_{-l}^{l} f_1(\tau)\sqrt{R(\tau)}\cot\frac{\tau - z_1}{a}d\tau. \quad (5.90)$$

By the similar way, we find that

$$\Psi_1(z_2) = \Psi_1^*(z_2) + \frac{\mu_1}{\mu_1 - \mu_2}\frac{c_0^*\tan\frac{z_1}{a} + c_1^*}{\sqrt{R(z_2)}} + \frac{\mu_1\beta^*}{\mu_1 - \mu_2}, \quad (5.91)$$

where

$$\frac{\mu_1 - \mu_2}{\mu_1}\Psi_1^*(z_2) = \frac{1}{2a\pi i}\int_{-l}^{l} g_1(\tau)\cot\frac{\tau - z_2}{a}d\tau + \frac{1}{2a\pi i\sqrt{R(z_2)}}\int_{-l}^{l} f_1(\tau)\sqrt{R(\tau)}\cot\frac{\tau - z_2}{a}d\tau, \quad (5.92)$$

in which c_0^*, c_1^* are undetermined real constants and β^*, complex.

In order to determine c_0, c_1, c_0^*, c_1^*, β and β^*, it is sufficient to consider the periodicity of displacements and the stresses at $z = \pm\infty i$.

First, consider the former:

$$[u + iv]_{\Lambda_\pm} = 0, \quad (5.93)$$

where $\Lambda_\pm$ is shown in Fig.2.6.

Substituting (3.5), (3.6), (5.89) and (5.91) into (5.93), and noting that

$$\frac{1}{\sqrt{R(\pm\infty i)}} = \pm\cos\frac{l}{a},$$

$$\int_{\Lambda_\pm}\frac{dz}{\sqrt{R(z_j)}} = \pm a\pi\cos\frac{l}{a},$$

$$\int_{\Lambda_\pm}\frac{\tan\frac{z_j}{a}}{\sqrt{R(z_j)}}dz = \int_{\Lambda_\pm}\frac{\cot\frac{\tau - z_j}{a}}{\sqrt{R(z_j)}}dz = a\pi i\cos\frac{l}{a}$$

$$(z_j = x + \mu_j y,\ j = 1,\ 2).$$

Thus, the above condition becomes

$$\mathrm{Re}\{\frac{\mu_2 p_1 - \mu_1 p_2}{\mu_2 - \mu_1}[\frac{1}{2}\int_{-l}^{l} g_1(\tau)d\tau + \frac{\cos\frac{l}{a}}{2}\int_{-l}^{l} f_1(\tau)\sqrt{R(\tau)}d\tau]$$

$$+ \frac{\mu_2 p_1}{\mu_2 - \mu_1} a\pi \cos\frac{l}{a}(c_0 i + c_2) + \frac{\mu_1 p_2}{\mu_1 - \mu_2} a\pi \cos\frac{l}{a}(c_0^* i + c_1^*)$$
$$+ a\pi p_1 \beta + a\pi p_2 \beta^* \} = 0, \tag{5.94}$$

$$\mathrm{Re}\{\frac{\mu_2 q_1 - \mu_1 q_2}{\mu_2 - \mu_1}[\frac{1}{2}\int_{-l}^{l} g_1(\tau)d\tau + \frac{\cos\frac{l}{a}}{2}\int_{-l}^{l} f_1(\tau)\sqrt{R(\tau)}\,d\tau]$$
$$+ \frac{\mu_2 q_1}{\mu_2 - \mu_1} a\pi \cos\frac{l}{a}(c_0 i + c_1) + \frac{\mu_1 q_2}{\mu_1 - \mu_2} a\pi \cos\frac{l}{a}(c_0^* i + c_1^*)$$
$$+ a\pi q_1 \beta + a\pi q_2 \beta^* \} = 0, \tag{5.95}$$

$$\mathrm{Re}\{\frac{\mu_2 p_1 - \mu_1 p_2}{\mu_2 - \mu_1}[-\frac{1}{2}\int_{-l}^{l} g_1(\tau)d\tau + \frac{\cos\frac{l}{a}}{2}\int_{-l}^{l} f_1(\tau)\sqrt{R(\tau)}\,d\tau]$$
$$+ \frac{\mu_2 p_1}{\mu_2 - \mu_1} a\pi \cos\frac{l}{a}(c_0 i - c_1) + \frac{\mu_1 p_2}{\mu_1 - \mu_2} a\pi \cos\frac{l}{a}(c_0^* i - c_1^*)$$
$$+ a\pi p_1 \beta + a\pi p_2 \beta^* \} = 0, \tag{5.96}$$

$$\mathrm{Re}\{\frac{\mu_2 q_1 - \mu_1 q_2}{\mu_2 - \mu_1}[-\frac{1}{2}\int_{-l}^{l} g_1(\tau)d\tau + \frac{\cos\frac{l}{a}}{2}\int_{-l}^{l} f_1(\tau)\sqrt{R(\tau)}\,d\tau]$$
$$+ \frac{\mu_2 q_1}{\mu_2 - \mu_1} a\pi \cos\frac{l}{a}(c_0^* i - c_1^*) + a\pi q_1 \beta + a\pi q_2 \beta^* \} = 0, \tag{5.97}$$

Next, consider the stresses at $z = \pm\infty i$:

$$\sigma_y(\pm\infty i) = \sigma_{xy}(\pm\infty i) = 0. \tag{5.98}$$

By substituting (3.3), (3.4), (5.89) and (5.91) into (5.98), these conditions are reduced to :

$$\mathrm{Re}\{-\frac{1}{2a\pi}\int_{-l}^{l} g_1(\tau)d\tau + \frac{\cos\frac{l}{a}}{2a\pi}\int_{-l}^{l} f_1(\tau)\sqrt{R(\tau)}\,d\tau$$
$$- \frac{\mu_2}{\mu_2 - \mu_1}(-c_0 i + c_1)\cos\frac{l}{a} + \beta$$
$$- \frac{\mu_1}{\mu_1 - \mu_2}(-c_0^* i + c_1^*)\cos\frac{l}{a} + \beta^* \} = 0, \tag{5.99}$$

$$\mathrm{Re}\{\frac{1}{2a\pi}\int_{-l}^{l} g_1(\tau)d\tau + \frac{\cos\frac{l}{a}}{2a\pi}\int_{-l}^{l} f_1(\tau)\sqrt{R(\tau)}\,d\tau$$
$$+ \frac{\mu_2}{\mu_2 - \mu_1}(c_0 i + c_1)\cos\frac{l}{a} + \beta$$
$$+ \frac{\mu_1}{\mu_1 - \mu_2}(c_0^* i + c_1^*)\cos\frac{l}{a} + \beta^* \} = 0, \tag{5.100}$$

$$\mathrm{Re}\{\frac{\mu_2 \mu_1}{\mu_2 - \mu_1}[-\cos\frac{l}{a}(-c_0 i + c_1) + \cos\frac{l}{a}(-c_0^* i + c_1^*)]$$

$$+ \mu_1\beta + \mu_2\beta^* \} = 0, \tag{5.101}$$

$$\mathrm{Re}\{ \frac{\mu_2\mu_1}{\mu_2 - \mu_1}[\cos\frac{l}{a}(c_0 i + c_1) - \cos\frac{l}{a}(c_0^* i + c_1^*)]$$
$$+ \mu_1\beta + \mu_2\beta^* \} = 0. \tag{5.102}$$

The constants c_0, c_1, c_0^*, c_1^*, β and β^* (8 real constants in fact) may be determined by (5.94) – (5.97) and (5.99) – (5.102). Thus, our problem for the case of symmetric loads is completely solved.

Let us consider the special case where the tension loads subjected on the different sides of the cracks are equal in magnititude and opposite in direction. In this case, $g_1(\tau) = 0$. The solution takes the following simple form:

$$\frac{\mu_2 - \mu_1}{\mu_2}\Phi_1(z_1) = \frac{1}{2a\pi i\sqrt{R(z_1)}}\int_{-l}^{l} f_1(\tau)\sqrt{R(\tau)}\cot\frac{\tau - z_1}{a}d\tau$$
$$+ \frac{c_0\tan\frac{z_1}{a} + c_1}{\sqrt{R(z_1)}} + \beta, \tag{5.103}$$

$$\frac{\mu_1 - \mu_2}{\mu_1}\Psi_1(z_2) = \frac{1}{2a\pi i\sqrt{R(z_2)}}\int_{-l}^{l} f_1(\tau)\sqrt{R(\tau)}\cot\frac{\tau - z_2}{a}d\tau$$
$$+ \frac{c_0^*\tan\frac{z_2}{a} + c_1^*}{\sqrt{R(z_2)}} + \beta^*, \tag{5.104}$$

where c_0, c_1, c_0^*, c_1^*, β, β^* may be defined by (5.94) – (5.97) and (5.99) – (5.102) as before, provided that $g_1(\tau)$ is replaced by zero.

For the subcase $g_1(\tau) = 0$ and $f_1(\tau) = -p$ ($p > 0$, a constant), the solution is much simpler:

$$\frac{\mu_2 - \mu_1}{\mu_2}\Phi_1(z_1) = -\frac{p}{2a\pi i\sqrt{R(z_1)}}\int_{-l}^{l} \sqrt{R(\tau)}\cot\frac{\tau - z_1}{a}d\tau$$
$$+ \frac{c_0\tan\frac{z_1}{a} + c_1}{\sqrt{R(z_1)}} + \beta, \tag{5.105}$$

$$\frac{\mu_1 - \mu_2}{\mu_1}\Psi_1(z_2) = -\frac{p}{2a\pi i\sqrt{R(z_2)}}\int_{-l}^{l} \sqrt{R(\tau)}\cot\frac{\tau - z_2}{a}d\tau$$
$$+ \frac{c_0^*\tan\frac{z_2}{a} + c_1^*}{\sqrt{R(z_2)}} + \beta^*. \tag{5.106}$$

In the mean time, in the equations for determining the undetermined constants, $g_1(\tau)$ and
$f_1(\tau)$ should be replaced by 0 and $-p$ respectively.

3. Anti-symmetric loads

Now, let us consider the problem of anti-symmetric loads (in the plane), i. e., $\tau_{xy}^{\pm}$ are given on the cracks. Denote the stress functions in this case by $\Phi_2(z_1)$ and $\Psi_2(z_2)$. By the condition of anti-symmetry,

$$\Phi_2(\tau) + \Psi_2(\tau) = 0, \ \tau \in L'.$$

Then, by (3.4), the boundary conditions become

$$\tau_{xy}^{+}(\tau) = (\mu_2 - \mu_1)\Phi_2^{+}(\tau) + (\bar{\mu}_2 - \bar{\mu}_1)\bar{\Phi}_2^{-}(\tau), \ \tau \in L_0,$$
$$\tau_{xy}^{-}(\tau) = (\mu_2 - \mu_1)\Phi_2^{-}(\tau) + (\bar{\mu}_2 - \bar{\mu}_1)\bar{\Phi}_2^{+}(\tau), \ \tau \in L_0,$$

or,

$$[(\mu_2 - \mu_1)\Phi_2(\tau) + (\bar{\mu}_2 - \bar{\mu}_1)\bar{\Phi}_2(\tau)]^{+} + [(\mu_2 - \mu_1)\Phi_2(\tau) + (\bar{\mu}_2 - \bar{\mu}_1)\bar{\Phi}_2(\tau)]^{-} = 2f_2(\tau), \ \tau \in L_0, \tag{5.107}$$

$$[(\mu_2 - \mu_1)\Phi_2(\tau) - (\bar{\mu}_2 - \bar{\mu}_1)\bar{\Phi}_2(\tau)]^{+} - [(\mu_2 - \mu_1)\Phi_2(\tau) - (\bar{\mu}_2 - \bar{\mu}_1)\bar{\Phi}_2(\tau)]^{-} = 2g_2(\tau), \ \tau \in L_0, \tag{5.108}$$

where

$$f_2(\tau) = \frac{1}{2}(\tau_{xy}^{+} + \tau_{xy}^{-}), \ g_2(\tau) = \frac{1}{2}(\tau_{xy}^{+} - \tau_{xy}^{-})$$

are periodic, assumed $\in H$.

Similar to the method in the previous paragraph, the general solution of (5.107) and (5.108) respectively are

$$(\mu_2 - \mu_1)\Phi_2(z_1) + (\bar{\mu}_2 - \bar{\mu}_1)\bar{\Phi}_2(z_1)$$
$$= \frac{1}{a\pi i\sqrt{R(z_1)}}\int_{-l}^{l} f_2(\tau)\sqrt{R(\tau)}\cot\frac{\tau - z_1}{a}d\tau + \frac{2d_0\tan\frac{z_1}{a} + 2d_1}{\sqrt{R(z_1)}},$$

$$(\mu_2 - \mu_1)\Phi_2(z_1) - (\bar{\mu}_2 - \bar{\mu}_1)\bar{\Phi}_2(z_1) = \frac{1}{a\pi i}\int_{-l}^{l} g_2(\tau)\cot\frac{\tau - z_1}{a}d\tau + 2\gamma,$$

where d_0, d_1 are undetermined real constants and γ, complex.

Thus, we may write

$$\Phi_2(z_1) = \bar{\Phi}_2^{*}(z_1) + \frac{1}{\mu_2 - \mu_1}\frac{d_0\tan\frac{z_1}{a} + d_1}{\sqrt{R(z_1)}} + \frac{\gamma}{\mu_2 - \mu_1}, \tag{5.109}$$

where we have put

$$(\mu_2 - \mu_1)\Phi_2^{*}(z_1) = \frac{1}{2a\pi i}\int_{-l}^{l} g_2(\tau)\cot\frac{\tau - z_1}{a}d\tau + \frac{1}{2a\pi i\sqrt{R(z_1)}}\int_{-l}^{l} f_2(\tau)\cot\frac{\tau - z_1}{a}d\tau.$$

Analogously, we find

$$\Psi_2(z_2) = \Psi_2^*(z_2) + \frac{1}{\mu_1 - \mu_2} \frac{d_0^* \tan \frac{z_2}{a} + d_1^*}{\sqrt{R(z_2)}} + \frac{\gamma^*}{\mu_1 - \mu_2}, \quad (5.110)$$

where d_0^*, d_1^* are undetermined real constants and γ^*, complex, and at the same time, we have defined

$$(\mu_1 - \mu_2)\Psi_2^*(z_2) = \frac{1}{2a\pi i}\int_{-l}^{l} g_2(\tau)\cot\frac{\tau - z_2}{a}d\tau + \frac{1}{2a\pi i\sqrt{R(z_2)}}\int_{-l}^{l} f_2(\tau)\sqrt{R(\tau)}\cot\frac{\tau - z_2}{a}d\tau.$$

As in the previous paragraph, by applying the periodicity of displacements and the condition at $z = \pm\infty i$, d_0^*, d_1^*, γ, γ^* may be determined by the following 8 equations:

$$\begin{aligned}\mathrm{Re}\{&\frac{p_1 - p_2}{\mu_2 - \mu_1}[\frac{1}{2}\int_{-l}^{l} g_2(\tau)d\tau + \frac{\cos\frac{l}{a}}{2}\int_{-l}^{l} f_2(\tau)\sqrt{R(\tau)}d\tau] \\ &+ \frac{p_1}{\mu_2 - \mu_1}a\pi\cos\frac{l}{a}(d_0 i + d_1) + \frac{p_2}{\mu_1 - \mu_2}a\pi\cos\frac{l}{a}(d_0^* i + d_1^*) \\ &+ a\pi p_1\gamma + a\pi p_2\gamma^*\} = 0, \end{aligned} \quad (5.111)$$

$$\begin{aligned}\mathrm{Re}\{&\frac{q_1 - q_2}{\mu_2 - \mu_1}[\frac{1}{2}\int_{-l}^{l} g_2(\tau)d\tau + \frac{\cos\frac{l}{a}}{2}\int_{-l}^{l} f_2(\tau)\sqrt{R(\tau)}d\tau] \\ &+ \frac{q_1}{\mu_2 - \mu_1}a\pi\cos\frac{l}{a}(d_0 i + d_1) + \frac{q_2}{\mu_1 - \mu_2}a\pi\cos\frac{l}{a}(d_0^* i + d_1^*) \\ &+ a\pi q_1\gamma + a\pi q_2\gamma^*\} = 0, \end{aligned} \quad (5.112)$$

$$\begin{aligned}\mathrm{Re}\{&\frac{p_1 - p_2}{\mu_1 - \mu_2}[-\frac{1}{2}\int_{-l}^{l} g_2(\tau)d\tau + \frac{\cos\frac{l}{a}}{2}\int_{-l}^{l} f_2(\tau)\sqrt{R(\tau)}d\tau] \\ &+ \frac{p_1}{\mu_2 - \mu_1}a\pi\cos\frac{l}{a}(d_0 i + d_1) + \frac{p_2}{\mu_1 - \mu_2}a\pi\cos\frac{l}{a}(d_0^* i - d_1^*) \\ &+ a\pi p_1\gamma + a\pi p_2\gamma^*\} = 0, \end{aligned} \quad (5.113)$$

$$\begin{aligned}\mathrm{Re}\{&\frac{q_1 - q_2}{\mu_2 - \mu_1}[-\frac{1}{2}\int_{-l}^{l} g_2(\tau)d\tau + \frac{\cos\frac{l}{a}}{2}\int_{-l}^{l} f_2(\tau)\sqrt{R(\tau)}d\tau] \\ &+ \frac{q_1}{\mu_2 - \mu_1}a\pi\cos\frac{l}{a}(d_0 i - d_1) + \frac{q_2}{\mu_1 - \mu_2}a\pi\cos\frac{l}{a}(d_0^* i - d_1^*) \\ &+ a\pi q_1\gamma + a\pi q_2\gamma^*\} = 0, \end{aligned} \quad (5.114)$$

$$\mathrm{Re}\{-\frac{1}{2a\pi}\int_{-l}^{l}g_2(\tau)d\tau+\frac{\cos\frac{l}{a}}{2a\pi}\int_{-l}^{l}f_2(\tau)\sqrt{R(\tau)}d\tau]-\frac{1}{\mu_2-\mu_1}(-d_0 i$$
$$+d_1)\cos\frac{l}{a}+\gamma-\frac{1}{\mu_1-\mu_2}(-d_0^* i+d_1)\cos\frac{l}{a}+\gamma^*\}=0, \quad (5.115)$$

$$\mathrm{Re}\{\frac{1}{2a\pi}\int_{-l}^{l}g_2(\tau)d\tau+\frac{\cos\frac{l}{a}}{2a\pi}\int_{-l}^{l}f_2(\tau)\sqrt{R(\tau)}d\tau]+\frac{1}{\mu_2-\mu_1}(d_0 i$$
$$+d_1)\cos\frac{l}{a}+\gamma+\frac{1}{\mu_1-\mu_2}(d_0^* i+d_1^*)\cos\frac{l}{a}+\gamma^*\}=0, \quad (5.116)$$

$$\mathrm{Re}\{\frac{1}{\mu_2-\mu_1}[-\cos\frac{l}{a}(-d_0 i+d_1)+\cos\frac{l}{a}(-d_0^* i+d_1^*)]+\gamma+\gamma^*\}=0, \quad (5.117)$$

$$\mathrm{Re}\{\frac{1}{\mu_2-\mu_1}[\cos\frac{l}{a}(d_0 i+d_1)-\cos\frac{l}{a}(d_0^* i+d_1^*)]+\gamma+\gamma^*\}=0. \quad (5.118)$$

Thus, our problem is completely solved.

In particular, consider the subcase $\overset{+}{\tau}_{xy}(\tau)=\overset{-}{\tau}_{xy}(\tau)$ or $g_2(\tau)=0$. Then, (5.109) and (5.110) are simplified respectively to

$$(\mu_2-\mu_1)\Phi_2(z_1)=\frac{1}{2a\pi i\sqrt{R(z_1)}}\int_{-l}^{l}f_2(\tau)\sqrt{R(\tau)}\cot\frac{\tau-z_1}{a}d\tau$$
$$+\frac{d_0\tan\frac{z_1}{a}+d_1}{\sqrt{R(z_1)}}+\gamma, \quad (5.119)$$
$$(\mu_2-\mu_1)\Psi_2(z_2)=\frac{1}{2a\pi i\sqrt{R(z_2)}}\int_{-l}^{l}f_2(\tau)\sqrt{R(\tau)}\cot\frac{\tau-z_2}{a}d\tau$$
$$+\frac{d_0^*\tan\frac{z_2}{a}+d_1^*}{\sqrt{R(z_2)}}+\gamma^*. \quad (5.120)$$

Next, consider the more special case where uniform shearing forces are applied on the cracks: $g_2(\tau)=0$ and $f_2(\tau)=-q$, $\tau\in L$. The solution becomes:

$$(\mu_2-\mu_1)\Phi_2(z_1)=\frac{-q}{2a\pi i\sqrt{R(z_1)}}\int_{-l}^{l}\sqrt{R(\tau)}\cot\frac{\tau-z_1}{a}d\tau$$
$$+\frac{d_0\tan\frac{z_1}{a}+d_1}{\sqrt{R(z_1)}}+\gamma, \quad (5.121)$$
$$(\mu_1-\mu_2)\Psi_2(z_2)=\frac{-q}{2a\pi i\sqrt{R(z_2)}}\int_{-l}^{l}\sqrt{R(\tau)}\cot\frac{\tau-z_2}{a}d\tau$$

$$+\frac{d_0^* \tan \frac{z_2}{a} + d_1^*}{\sqrt{R(z_2)}} + \gamma^* . \quad (5.122)$$

4. Stress intensity factors

In order to analyze the stress distribution arround the tips of L_0, put, in the neighborhood of $t = l$,

$$x = l + r\cos\theta, \; y = l + r\sin\theta,$$

where r/l is assumed to be sufficiently small and the polar coordinates r, θ represent respectively the radial distance of the point $z = x + iy$ to the tip $t = l$ of the crack and the angle of inclination of the radial ray to the crack.

Note that, when $z_j \approx l$,

$$(\tan^2 \frac{z_j}{a} - \tan^2 \frac{l}{a})^{\frac{1}{2}} \approx \sec^2 \frac{l}{a}[2r(\cos\theta + \mu_j \sin\theta)]^{\frac{1}{2}}, \; j = 1, 2.$$

Thus, the stress functions, either for the case of symmetric or anti-symmetric loads, may be written as

$$\left.\begin{aligned} \Phi_j(z_1) &= \frac{F_j}{[r(\cos\theta + \mu_j \sin\theta)]^{\frac{1}{2}}} + O(1), \\ \Psi_j(z_2) &= \frac{G_j}{[r(\cos\theta + \mu_j \sin\theta)]^{\frac{1}{2}}} + O(1), \end{aligned}\right\} \quad j = 1, 2, \quad (5.123)$$

where

$$F_1 = \frac{k_1 \mu_2 \cos \frac{l}{a}}{2\sqrt{2}(\mu_2 - \mu_1)}, \qquad F_2 = \frac{k_2 \cos \frac{l}{a}}{2\sqrt{2}(\mu_2 - \mu_1)},$$

$$G_1 = \frac{k_1 \mu_1 \cos \frac{l}{a}}{2\sqrt{2}(\mu_1 - \mu_2)}, \qquad G_2 = \frac{k_2 \cos \frac{l}{a}}{2\sqrt{2}(\mu_1 - \mu_2)} = -F_2,$$

in which k_j, $j = 1, 2$, are called the *stress intensity factors*, which may be evaluated directly from the stress functions $\Phi_j(z_1)$ or $\Psi_j(z_2)$, namely, by (5.103) and (5.121),

$$\left.\begin{aligned} k_1 &= 2\sqrt{2} \frac{\mu_2 - \mu_1}{\mu_2} \lim_{z_1 \to t_0} (\tan \frac{z_1}{a} - \tan \frac{t_0}{a})^{\frac{1}{2}} \Phi_1(z_1), \\ k_2 &= 2\sqrt{2}(\mu_2 - \mu_1) \lim_{z_1 \to t_0} (\tan \frac{z_1}{a} - \tan \frac{t_0}{a})^{\frac{1}{2}} \Phi_2(z_1), \end{aligned}\right\} \quad (5.124)$$

where $t_0 \in L_0$, or,

$$\left.\begin{aligned} k_1 &= 2\sqrt{2}\,\frac{\mu_1 - \mu_2}{\mu_1} \lim_{z_2 \to t_0} \left(\tan\frac{z_2}{a} - \tan\frac{t_0}{a}\right)^{\frac{1}{2}} \Psi_1(z_2), \\ k_2 &= 2\sqrt{2}(\mu_1 - \mu_2) \lim_{z_2 \to t_0} \left(\tan\frac{z_2}{a} - \tan\frac{z_0}{a}\right)^{\frac{1}{2}} \Psi_2(z_2). \end{aligned}\right\} \tag{5.125}$$

When $a \to \infty$, the stress intensity factors for the case where there is only one single crack on the x-axis may be obtained, which are identical to the result due to G.C. Sih and H. Liebowitz [1].

Chapter Ⅵ

Doubly-Periodic Problems in Plane Elasticity

In practical engineering, sometimes there occur doubly-periodic problems in plane elasticity. In comparison with non-periodic or simply periodic case, such problems are quite not developed completely, except in some very particular cases. For instance, for the case where there is only one single hole in the fundamental doubly-periodic parallelogram, W. T. Koiter [2] discussed the expressions of the complex stress functions and gave the formulation and solution of its first fundamental problem. However, when there are many holes in the parallelogram, the stress functions are multi-valued and much more complicated. Moreover, if there occur cracks instead of holes in it, the situation is quite different.

In the present chapter, we would give a brief sketch of the doubly periodic elastic problems for isotropic media in the general case, including the general expressions of stress functions and the general formulation of the doubly-periodic fundamental problems. These results were first exploited in Lu [4] and independently established later by Zheng [1].

§ 1. Preliminaries

1. General notion

Let ω_1 and ω_2 be two complex numbers with $\mathrm{Im}(\omega_2/\omega_1)\neq 0$. A function $F(z)$ is called *doubly-periodic* in certain domain D (not necessarily connected) with periods $2\omega_1$ and $2\omega_2$ if

$$F(z+2\omega_j)=F(z),\ z\in D,\ j=1,\ 2; \tag{6.1}$$

of course D itself must be doubly-periodic, i.e., $z\in D$ implies $z+2\omega_j\in D$. In the sequel, we always assume $\mathrm{Im}(\omega_2/\omega_1)>0$, as othewise we need only replace, for example, ω_2 by $-\omega_2$. Let both ω_1 and ω_2 be fixed all over the chapter.

Any parallelogram with vertices z_0, $z_0+2\omega_1$, $z_0+2\omega_1+2\omega_2$, $z_0+2\omega_2$ (no doubt in counter-clockwise order) for any z_0 is called a *doubly-periodic parallelogram*. In particular, that one with $z_0=0$ is called the fundamental doubly-periodic parallelogram, or simply, the *fundamental cell*, denoted by P_0. Let the boundary contour Γ_0 of P_0 be oriented counter-clockwisely and its four sides be denoted by Γ_j, $j=1, 2, 3, 4$, as shown in Fig. 6.1.

The point $\Omega_{mn} = 2m\omega_1 + 2n\omega_2$ ($m, n = 0, \pm 1, \pm 2, \cdots$) is congruent to 0 (mod $2\omega_j$), and in general, $\Omega_{mn} + z \equiv z$ (mod $2\omega_j$).

A function $F(z)$ satisfying, instead of (6.1),

$$F(z+2\omega_j) = F(z) + \alpha_k, \quad z \in D, \quad k = 1, 2, \tag{6.2}$$

is called *doubly quasi-periodic* in D, where α_1 and α_2 are two constants, called the *addenda* of $F(z)$.

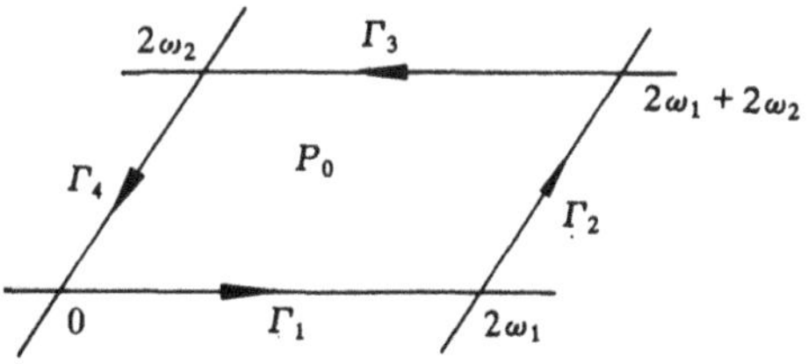

Fig. 6.1

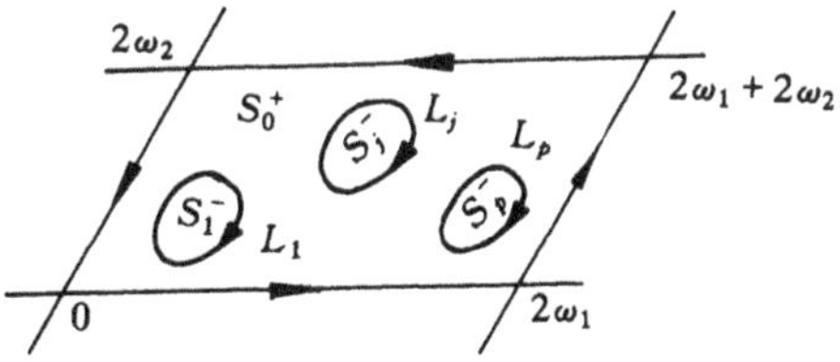

Fig. 6.2

Assume there is a finite number of smooth closed contours $L_1, L_2, \cdots, L_p$ in P_0, oriented counter-clockwisely. The interior region enclosed by L_j is denoted by S_j^- (Fig. 6.2). Let $S_0^- = \sum_{j=1}^{p} S_j^-$, $S_0^+ = \overline{P}_0 \setminus \overline{S}_0^-$, and S^+, S^- be the unions of the domains congruent to S_0^+ and S_0^- respectively. Thus, S^+ is in fact a doubly periodic connected region. Moreover, denote $L_0 = \sum_{j=1}^{p} L_j$ and L, the union of contours congruent to L_0.

A doubly-periodic function $\Phi(z)$, holomorphic in S^+ and S^-, and continuous to L on both its sides, is called *sectionally doubly-periodic holomorphic*.

In general, $z = \infty$ is an essential singular point of $\Phi(z)$. However, "the point at infinity" plays no role in doubly-periodic problems.

2. Weierstrass functions

We recall some basic definitions and properties of Weierstrass functions, related to elliptic functions, to be used in the sequel, which may be found in many textbooks on complex analysis, e.g., Ahlfors [1].

A meromorphic function (i.e., a single-valued analytic function in the complex plane with poles only as its singularities) with periods $2\omega_1$ and $2\omega_2$ is called an *elliptic function*.

The function

$$\zeta(z) = \frac{1}{z} + \sum_{m,n}{}'\left(\frac{1}{z - \Omega_{mn}} + \frac{1}{\Omega_{mn}} + \frac{z}{\Omega_{mn}^2}\right), \tag{6.3}$$

by definition, is called the *Weierstrass ζ-function*, where $\sum_{m,n}'$ denotes the summation over all integers m, n except $m = n = 0$. It is a meromorphic function possessing simple poles Ω_{mn}, m, $n = 0, \pm 1, \pm 2, \cdots$, with residue 1 and is an odd function evidently. Of course it is not doubly periodic but doubly quasi-periodic:

$$\zeta(z + 2\omega_k) = \zeta(z) + 2\eta_k,\ k = 1, 2, \tag{6.4}$$

where $\eta_k = \zeta(\omega_k)$ satisfying

$$2\omega_2\eta_1 - 2\omega_1\eta_2 = \pi i, \tag{6.5}$$

which may be easily verified by the residue theorem applied to the integral $\frac{1}{2\pi i}\int_{\Gamma_0}\zeta(z)dz$. ①

By (6.4), it is seen that $\zeta'(z)$ is doubly-periodic. We define an even function

$$\mathscr{P}(z) = -\zeta'(z) = \frac{1}{z^2} + \sum_{m,n}{}'\left\{\frac{1}{(z - \Omega_{mn})^2} - \frac{1}{\Omega_{mn}^2}\right\}, \tag{6.6}$$

called the *Weierstrass ℘-function*, which is an elliptic function with a unique double pole at $z = 0$ in P_0. By repeated differentiation, we get $\mathscr{P}^{(k)}(z) = -\zeta^{(k+1)}(z)$ $(k = 1, 2, \cdots)$ is an elliptic function with unique pole $z = 0$ of order $k + 2$ in P_0.

Another function $\sigma(z)$ will be used later. Considering $e^{\int\zeta(z)dz}$, we define an odd function, called the *Weiertrass σ-function*,

$$\sigma(z) = z\prod_{m,n}{}'\left(1 - \frac{z}{\Omega_{mn}}\right)\exp\left\{\frac{1}{\Omega_{mn}} + \frac{z^2}{2\Omega_{mn}^2}\right\}, \tag{6.7}$$

where $\prod_{m,n}'$ denotes the multiplication over all m, n except $m = n = 0$. Here, in the infinite product, the factor $z - \Omega_{mn}$ arisen in $e^{\int\zeta(z)dz}$ is replaced by $1 - z/\Omega_{mn}$ so as to guarantee its convergence. It takes $z = \Omega_{mn}$ $(m, n = 0, \pm 1, \pm 2, \cdots)$ as its simple zero-point. It is neither doubly-periodic nor doubly quasi-periodic but fulfilling the following equalities:

$$\sigma(z + 2\omega_k) = -\sigma(z)e^{2\eta_k(z+\omega_k)},\ k = 1, 2. \tag{6.8}$$

In fact, it is evident that

$$\sigma'(z)/\sigma(z) = \zeta(z). \tag{6.9}$$

---

① If $\mathrm{Im}(\omega_2/\omega_1) < 0$, then the right-hand member of (6.5) should be replaced by $-\pi i$.

By (6.4), we then have

$$\frac{\sigma'(z+2\omega_k)}{\sigma(z+2\omega_k)} = \frac{\sigma'(z)}{\sigma(z)} + 2\eta_k, \quad k = 1, 2,$$

which, by integration, immediately follow that

$$\sigma(z+2\omega_k) = C_k\sigma(z)e^{2\eta_k z} \quad (k = 1, 2).$$

Putting $z = -\omega_k$ in this equation and noting that $\sigma(z)$ is odd, we readily have $C_k = -e^{2\eta_k\omega_k}$ which follows (6.8).

§ 2. General Expressions of Complex Stress Functions

1. General explanations

In doubly-periodic problems of plane elasticity, we consider the following case: the elastic region occupies a doubly-periodic connected region S and the stresses (including the boundary external stresses) are also doubly-periodic:

$$\begin{aligned} &\sigma_x(z+2\omega_k) = \sigma_x(z), \ \sigma_y(z+2\omega_k) = \sigma_y(z), \\ &\tau_{xy}(z+2\omega_k) = \tau_{xy}(z), \ z \in S, \ k = 1, 2. \end{aligned} \tag{6.10}$$

By the generalized Hooke's law, the displacement function $g(z) = u(z) + iv(z)$ must be doubly quasi-periodic:

$$g(z+2\omega_k) = g(z) + g_k, \quad k = 1, 2, \tag{6.11}$$

where g_1 and g_2 are constants, the addenda of $g(z)$. A rigid translation of the whole elastic body does not change the addenda. A rigid rotation about the origin with displacement $i\varepsilon z$ ($\varepsilon > 0$ is a small positive number, Cf. Lu [6]) possesses the increments $2i\omega_k$, $k = 1, 2$, of the addends, which makes no influence on the relative displacements among the points of the elastic body as well as the stress distribution.

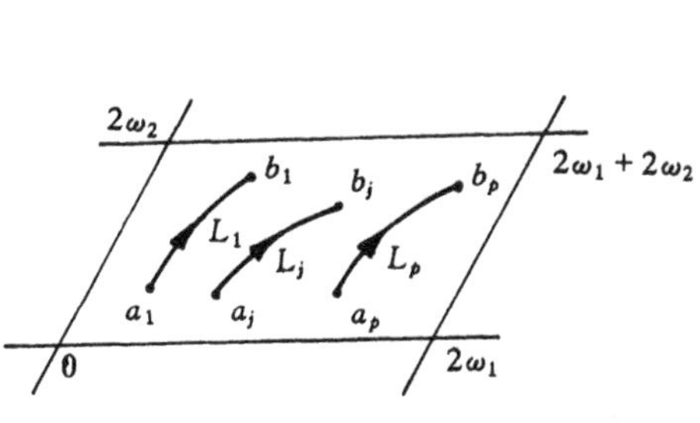

Fig.6.3

Let the elastic body occupy the region $S = S^+$ as shown in Fig.6.2, or a region S similar to S^+ but the holes $L_1, \cdots, L_p$ in P_0 are replaced by non-intersecting smooth cracks $L_j = a_jb_j$, $j = 1, \cdots, p$, oriented from a_j to b_j (Fig. 6. 3). Denote the principal vector of the stresses on the boundary contour L_j of the hole S_j^- or the resultant of the principal vectors of the external stresses on the two sides of the crack $L_j = a_jb_j$ by $X_j + iY_j$. Since the principal vectors of the external stresses applied to the two pairs of the opposite sides of P_0 cancel each other respectively by double periodicity, we have, by equilibrium,

$$\sum_{j=1}^{p}(X_j + iY_j) = 0. \tag{6.12}$$

Therefore, if there occurs only one single hole or crack in P_0, then we must have
$X_1 + iY_1 = 0$.

We always assume S is of isotropic medium with elastic constants κ and μ. We would establish the general expressions of the complex stress functions for the case of holes and the case of cracks seperately.

2. The case of region with holes

Consider the case of doubly-periodic region $S = S^+$ with p holes $S_1^-, \cdots, S_p^-$ in P_0(Fig. 6.2).

Take an arbitrary point z_j in each S_j^- $(j = 1, \cdots, p)$. According to the general theory of plane elasticity, by double periodicity, the *complex stress functions* $\varphi(z)$ and $\psi(z)$ may be expressed as

$$\varphi(z) = -\frac{1}{2\pi(\kappa+1)}\sum_{j=1}^{p}(X_j + iY_j)\log\sigma(z - z_j) + \varphi_0(z), \tag{6.13}$$

$$\psi(z) = \frac{\kappa}{2\pi(\kappa+1)}\sum_{j=1}^{p}(X_j - iY_j)\log\sigma(z - z_j) + \psi_1(z), \tag{6.14}$$

$(z \in S^+)$, where $\varphi_0(z)$ and $\psi_1(z)$ are holomorphic in S^+, $\sigma(z)$ is the Weierstrass σ-function and the branch of logarithm may be taken arbitrarily. Thus, their multi-valued parts have been seperated out. By differentiation (write $\Phi(z) = \varphi'(z)$, $\Phi_0(z) = \varphi_0'(z)$, etc.), we have

$$\Phi(z) = -\frac{1}{2\pi(\kappa+1)}\sum_{j=1}^{p}(X_j + iY_j)\zeta(z - z_j) + \Phi_0(z), \tag{6.15}$$

$$\Psi(z) = \frac{\kappa}{2\pi(\kappa+1)}\sum_{j=1}^{p}(X_j - iY_j)\zeta(z - z_j) + \Psi_1(z), \tag{6.16}$$

where $\zeta(z) = \sigma'(z)/\sigma(z)$ is the Weierstrass ζ-function. By (6.12), it is readily seen that the summation parts in (6.15) and (6.16) are doubly-periodic.

We assure that $\Phi_0(z) = P + iQ$ is doubly-periodic. In fact, by double periodicity of the stresses and (2.1), we see at once that P is doubly-periodic and so do $\frac{\partial Q}{\partial x}$ and $\frac{\partial Q}{\partial y}$ by the Cauchy-Riemann equations. Thereby Q is doubly quasi-periodic: $Q(z+2\omega_k)=Q(z)+\nu_k$, $k=1, 2$, where ν_1, ν_2 are real constants, and then $\Phi_0(z+2\omega_k)=\Phi_0(z)+\nu_k$. Let

$$\Phi_*(z) = \Phi_0(z) - A_0 z - B_0\zeta(z-z_1), \tag{6.17}$$

where

$$A_0 = \frac{1}{\pi i}(\nu_2\eta_1 - \nu_1\eta_2),\ B_0 = \frac{1}{\pi i}(\omega_2\nu_1 - \omega_1\nu_2) \tag{6.18}$$

($\eta_k=\zeta(\omega_k)$, $k=1, 2$). We may verify directly that $\Phi_*(z)$ is doubly-periodic on account of (6.4) and (6.5), and holomorphic in S^+. Multiplying both sides of (6.17) by $\frac{dz}{2\pi i}$ and integrating along Γ_0, we immediately get $B_0=0$ since $\varphi_0(z)$ is already known to be single-valued. Therefore,

$$\Phi_0(z+2\omega_k) - \Phi_0(z) = 2A_0\omega_k = i\nu_k,\ k=1, 2.$$

If $A_0\neq 0$, then $2A_0=i\nu_k/\omega_k$ and so $\omega_2/\omega_1=\nu_2/\nu_1$ is real, which is a contradiction. Hence $A_0=0$, which means that $\Phi_0(z)$ is doubly-periodic.

Choose constants A and B such that the function

$$D(z) = Az + B\zeta(z-z_1) \tag{6.19}$$

holomorphic in S^+ is doubly quasi-periodic with addenda $2\bar{\omega}_k$:

$$D(z+2\omega_k) = D(z) + 2\bar{\omega}_k,\ k = 1, 2, \tag{6.20}$$

and so

$$m(z) = \bar{z} - D(z) \tag{6.21}$$

is doubly-periodic in S^+ (but not analytic). It is easy to verify that, by (6.5),

$$A = \frac{2}{\pi i}(\bar{\omega}_2\eta_1 - \bar{\omega}_1\eta_2),\ B = \frac{2}{\pi i}(\bar{\omega}_2\omega_1 - \bar{\omega}_1\omega_2), \tag{6.22}$$

which shows that B is real. Put

$$\Psi_2(z) = \Psi_1(z) + D(z)\Phi'(z) = \Psi_1(z) + \bar{z}\Phi'(z) - m(z)\Phi'(z) \tag{6.23}$$

which is doubly-periodic and holomorphic in S^+. Thus, (6.16) may be rewritten as

$$\Psi(z) = \frac{\kappa}{2\pi(\kappa+1)}\sum_{j=1}^{p}(X_j - iY_j)\zeta(z-z_j) - D(z)\Phi_0'(z) + \Psi_2(z). \tag{6.24}$$

(6.15) and (6.24) are the general expressions of $\Phi(z)$ and $\Psi(z)$ respectively, where

$\Phi_0(z)$ and $\Psi_2(z)$ are doubly-periodic holomorphic functions in S^+.

Returning to (6.14), we may write

$$\psi(z) = \frac{\kappa}{2\pi(\kappa+1)}\sum_{j=1}^{p}(X_j - iY_j)\log\sigma(z - z_j) - D(z)\Phi(z) + \psi_0(z), \tag{6.25}$$

where $\psi_0(z)$, as $\varphi_0(z)$, is determined up to a constant term, with

$$\psi_0'(z) = \Psi_0(z) = \Psi_2(z) + D'(z)\Phi(z) \tag{6.26}$$

to be doubly-periodic and holomorphic in S^+. (6.13) and (6.25) are the general expressions of $\varphi(z)$ and $\psi(z)$ respectively, in which $\varphi_0(z)$ and $\psi_0(z)$ are holomorphic in S^+ since the multi-valued parts of $\varphi(z)$ and $\psi(z)$ have been seperated out and are doubly quasi-periodic on account of $\varphi_0'(z) = \Phi_0(z)$ and $\psi_0'(z) = \Psi_0(z)$ to be doubly-periodic. Let

$$\varphi_0(z + 2\omega_k) = \varphi_0(z) + \varphi_k, \quad \psi_0(z + 2\omega_k) = \psi_0(z) + \psi_k, \quad k = 1, 2, \tag{6.27}$$

where φ_k and ψ_k are constants.

3. The case of region with cracks

Assume that there are p cracks $L_j = a_jb_j$, $j = 1, \cdots, p$ in P_0 (Fig.6.3). The principal vectors of the external stresses on the positive and the negative sides of L_j are denoted by $X_j^{\pm} + iY_j^{\pm}$ respectively, the resultant of which is

$$X_j + iY_j = (X_j^+ + iY_j^+) + (X_j^- + iY_j^-) \ (j = 1, \cdots, p) \tag{6.28}$$

with (6.12) fulfilled. If there is only one crack L_1 in P_0, then $X_1^+ = -X_1^-$, $Y_1^+ = -Y_1^-$.

In place of (6.13) and (6.14), we may write

$$\varphi(z) = -\frac{1}{4\pi(\kappa+1)}\sum_{j=1}^{p}(X_j + iY_j)\log\sigma(z - a_j)\sigma(z - b_j) + \varphi_0^*(z) \tag{6.29}$$

and

$$\psi(z) = \frac{\kappa}{4\pi(\kappa+1)}\sum_{j=1}^{p}(X_j - iY_j)\log\sigma(z - a_j)\sigma(z - b_j) + \psi_1^*(z) \tag{6.30}$$

($z \in S$), and in which $\varphi_0^*(z)$ and $\psi_1^*(z)$ are single-valued so that their multi-valued parts have been seperated out. However, since the summation parts in these expressions possess logarithmic singularities at the tips a_j, b_j of the crack L_j ($j = 1, \cdots, p$) and hence their derivatives have singularities there of order 1, which is not convenient in applications. To overcome this deficiency, we introduce functions

$$H_j(z) = \int_{L_k} h_j(t)\zeta(t-z)dt, \quad z \in S, \; j = 1, \cdots, p, \tag{6.31}$$

where

$$h_j(t) = \frac{2t - a_j - b_j}{b_j - a_j}. \tag{6.32}$$

Because the principal part of $\zeta(t-z)$ at $t=z$ is $\dfrac{1}{t-z}$ and $h_j(a_j)=1$, $h_j(b_j)=-1$, so we know that $H_j(z)$ possesses logarithmic singularities at $z=a_j$ and b_j as $\log\sigma(z-a_j)\sigma(z-b_j)$ (Cf. Muskhelishvili [1]) and hence $\log\sigma(z-a_j)\sigma(z-b_j) - H_j(z)$ are bounded there. Thus, (6.29) and (6.30) may be respectively written as

$$\varphi(z) = -\frac{1}{4\pi(\kappa+1)}\sum_{j=1}^{p}(X_j + iY_j)[\log\sigma(z-a_j)\sigma(z-b_j) - H_j(z)] + \varphi_0(z), \tag{6.33}$$

$$\psi(z) = \frac{\kappa}{4\pi(\kappa+1)}\sum_{j=1}^{p}(X_j - iY_j)[\log\sigma(z-a_j)\sigma(z-b_j) - H_j(z)] + \psi_1(z), \tag{6.34}$$

where both $\varphi_0(z)$ and $\psi_1(z)$ are still holomorphic is S and have singularities at all the tips of the cracks of the same type as $\varphi(z)$ and $\psi(z)$ respectively. By the way, we mention that
$H_j(z)$ is doubly-periodic since $\int_{a_j}^{b_j} h_j(t)dt = 0$.

By differentiation, we have

$$\Phi(z) = -\frac{1}{4\pi(\kappa+1)}\sum_{j=1}^{p}(X_j + iY_j)[\zeta(z-a_j) + \zeta(z-b_j) - H_j'(z)] + \Phi_0(z), \tag{6.35}$$

$$\Psi(z) = \frac{\kappa}{4\pi(\kappa+1)}\sum_{j=1}^{p}(X_j - iY_j)[\zeta(z-a_j) + \zeta(z-b_j) - H_j'(z)] + \Psi_1(z), \tag{6.36}$$

where

$$H_j'(z) = \int_{L_j} h_j(t)\mathscr{P}(t-z)dt \tag{6.37}$$

is of course doubly-periodic.

As in the previous paragraph, it may be easily verified that $\Phi_0(z)$ is doubly-periodic. On considering $\Psi_1(z)$, put

$$\zeta_1(z) = \frac{6}{(b_1 - a_1)^3}\int_{L_1}(b_1 - t)(t - a_1)\zeta(t-z)dt, \tag{6.38}$$

which is holomorphic in S, bounded at a_j, b_j $(j=1, \cdots, p)$, and

$$\zeta_1(z + 2\omega_k) = \zeta_1(z) + 2\eta_k, \ k = 1, 2, \tag{6.39}$$

that is, $\zeta_1(z)$ have the addends as the same as $\zeta(z)$. in place of $D(z)$ in (6.19), we define

$$D_1(z) = Az + B\zeta_1(z), \tag{6.40}$$

where A, B are still given by (6.22). Thus, $D_1(z)$ also has the addends $2\bar{\omega}_1$, $2\bar{\omega}_2$ and so

$$m_1(z) = \bar{z} - D_1(z) \tag{6.41}$$

is doubly-periodic. Therefore, in place of (6.34), we may write

$$\begin{aligned}\Psi(z) = {} & \frac{\kappa}{4\pi(\kappa+1)}\sum_{j=1}^{p}(X_j - iY_j)[\zeta(z-a_j) + \zeta(z-b_j) - H_j'(z)] \\ & - D_1(z)\Phi'(z) + \Psi_2(z),\end{aligned} \tag{6.42}$$

where $\Psi_2(z)$ is already single-valued and doubly-periodic in S. Since $D_1(z)$ keeps bounded near all the tips of the cracks, its appearing in (6.42) makes no influence on the behaviour of singularities of $\Psi(z)$ or $\Psi_2(z)$ as $D(z)$ before. (6.35) and (6.42) are the general expressions of $\Phi(z)$ and $\Psi(z)$ respectively in the case of cracks, where both $\Phi_0(z)$ and $\Psi_2(z)$ are doubly-periodic holomorphic functions.

By returning to $\psi(z)$, (6.34) may be rewritten as

$$\begin{aligned}\psi(z) = {} & \frac{\kappa}{4\pi(\kappa+1)}\sum_{j=1}^{p}(X_j - iY_j)[\log\sigma(z-a_j)(z-b_j) - H_j(z)] \\ & - D_1(z)\Phi(z) + \psi_0(z).\end{aligned} \tag{6.43}$$

At this time, (6.26) remains valid when $D(z)$ is replaced by $D_1(z)$. (6.33) and (6.43) are the general expressions of the stress function in our case. The addenda of $\varphi(z)$ and $\psi(z)$ are again denoted by φ_k, ψ_k $(k=1, 2)$ respectively.

§ 3. Doubly-periodic Fundamental Problems

1. Relations among the addenda of related functions

Before discussing the doubly-periodic fundamental problems, we would clarify the relations among the addenda φ_k, ψ_k and g_k $(k=1, 2)$ of the functions $\varphi_0(z)$, $\psi_0(z)$ appeared in the expressions of stress functions and the displacement function $g(z)$ respectively.

First, we introduce some mechanical quantities. Let the external stresses on Γ_0 be
$X_n(\tau) + iY_n(\tau)$ $(\tau \in \Gamma_0)$. Then, the principal vector of the external stresses applied on the side Γ_k is

$$F_k = \int_{\Gamma_k} [X_n(\tau) + iY_n(\tau)] d\sigma \quad (k = 1, 2, 3, 4), \tag{6.44}$$

where σ is the arc-length parameter on Γ. By the condition of double periodicity, $F_3 = -F_1$, $F_4 = -F_2$. Set

$$f(\tau) = i\int_0^\tau [X_n(t) + iY_n(t)] d\sigma, \ \tau \in \Gamma_0, \tag{6.45}$$

which is single-valued since the resultant of the principal vectors on the whole Γ_0 is zero. It is evident that

$$F_k = -i[f(\tau)]_{\Gamma_k} \tag{6.46}$$

where $[f(\tau)]_{\Gamma_k}$ represents the increment of $f(\tau)$ when τ describes along Γ_k in its positive direction, that means,

$$[f(\tau)]_{\Gamma_1} = f(2\omega_1) - f(0) = f(2\omega_1), \ [f(\tau)]_{\Gamma_2} = f(2\omega_1 + 2\omega_2) - f(2\omega_1).$$

It is known that, by (2.35),

$$\varphi(\tau) + \tau\overline{\varphi'(\tau)} + \overline{\psi(\tau)} = f(\tau) + \text{const}, \ \tau \in \Gamma_0, \tag{6.47}$$

and so

$$[\varphi(\tau) + \tau\overline{\varphi'(\tau)} + \overline{\psi(\tau)}]_{\Gamma_k} = iF_k, \ k = 1, 2, 3, 4. \tag{6.48}$$

Consider the case of holes. Using the property (6.8) of $\sigma(z)$:

$$\sigma(z - z_j + 2\omega_k) = -e^{2\eta_k(z - z_j + \omega_k)}\sigma(z - z_j), \ k = 1, 2; \ j = 1, \cdots, p, \tag{6.49}$$

we have, by substituting it into (6.13), (6.25) and noting (6.12),

$$\frac{1}{\pi(\kappa+1)}\sum_{j=1}^{p}(X_j + iY_j)(\eta_k z_j - \kappa\bar{\eta}_k\bar{z}_j) + \varphi_k + \bar{\psi}_k = iF_k, \ k = 1, 2. \tag{6.50}$$

On the other hand, by similar substitution into (2.3), we obtain

$$\frac{2\kappa}{\pi(\kappa+1)}\sum_{j=1}^{p}(X_j + iY_j)\text{Re}\{\eta_k z_j\} + \kappa\varphi_k - \bar{\psi}_k = 2\mu g_k, \ k = 1, 2, \tag{6.51}$$

on account of $[g(\tau)]_{\Gamma_k} = g_k$ $(k = 1, 2)$.

Then, by (6.50) and (6.51), we readily get

$$\left.\begin{aligned}\varphi_k &= -\frac{\eta_k}{\pi(\kappa+1)}\sum_{j=1}^{p}(X_j+iY_j)z_j+\frac{1}{\kappa+1}(iF_k+2\mu g_k),\\ \psi_k &= \frac{\eta_k}{\pi(\kappa+1)}\sum_{j=1}^{p}(X_j-iY_j)z_j-\frac{1}{\kappa+1}(i\kappa\bar{F}_k+2\bar{\eta}_k),\end{aligned}\right\}\quad k=1,\ 2. \tag{6.52}$$

Moreover, let us consider the principal moment M_0 of the external stresses on Γ_0. Obviously, its corresponding part on $\Gamma_1+\Gamma_2$ is $\frac{1}{2}M_0$. Evidently, the principal moment of the external stresses applied on L_0 is $M_{L_0}=-M_0$.

We would find the relation between M_0 and F_k. Let $\tau=\xi+i\eta$. Then,

$$M_0=\int_{\Gamma_0}[\xi Y_n(\tau)-\eta X_n(\tau)]d\sigma=\mathrm{Im}\int_{\Gamma_0}\bar{\tau}[X_n(\tau)+iY_n(\tau)]d\sigma.$$

On the other hand, by the double periodicity of the stresses, we know that $X_n(\tau+2\omega_k)=$
$-X_n(\tau)$, $Y_n(\tau+2\omega_k)=-Y_n(\tau)$, $\tau\in\Gamma_k$ $(k=1,\ 2)$. Consequently, we obtain by the above equality,

$$M_0=-M_{L_0}=2\mathrm{Im}\{\bar{\omega}_1F_2-\bar{\omega}_2F_1\}. \tag{6.53}$$

In the mean time, we may find the relation between M_0 and ψ_k. Since (Cf. Muskhelishvili [1] or Lu [6])

$$M_0=-\mathrm{Re}\int_{\Gamma_0}\tau\psi(\tau)d\tau,$$

we get at once, by the double periodicity of $\Psi(\tau)$,

$$M_0=2\mathrm{Re}\{\omega_2[\psi(\tau)]_{\Gamma_1}-\omega_1[\psi(\tau)]_{\Gamma_2}\},$$

and, by substitutoin from (6.25) and using (6.12), it is observed that

$$\begin{aligned}M_0=2\mathrm{Re}\{&\omega_2[\frac{\kappa\eta_1}{\pi(\kappa+1)}\sum_{j=1}^{p}(X_j-iY_j)z_j-2\bar{\omega}_1\Phi(0)+\psi_1]\\&-\omega_1[\frac{\kappa\eta_2}{\pi(\kappa+1)}\sum_{j=1}^{p}(X_j-iY_j)z_j-2\bar{\omega}_2\Phi(0)+\psi_2]\},\end{aligned}$$

which follows, by (6.5) and (6.15), after simplification,

$$\begin{aligned}M_0=&-\frac{\kappa}{\kappa+1}\mathrm{Im}\{\sum_{j=1}^{p}(X_j-iY_j)z_j\\&+\frac{2}{\pi(\kappa+1)}\mathrm{Re}\{(\omega_1\bar{\omega}_2-\bar{\omega}_1\omega_2)\sum_{j=1}^{p}(X_j+iY_j)\zeta(z_j)\}\\&+2\mathrm{Re}\{\omega_2\psi_1-\omega_1\psi_2\}+4\mathrm{Re}\{(\omega_1\bar{\omega}_2-\bar{\omega}_1\omega_2)\Phi_0(0)\}.\end{aligned} \tag{6.54}$$

Since $\varphi_0(z)$ may be changed by a term $i\varepsilon z$ (ε: real), we may always assume

$$\operatorname{Im}\varphi_0'(0) = \operatorname{Im}\Phi_0(0) = 0, \tag{6.55}$$

so that the last term on the right side of (6.54) may be omitted:

$$\begin{aligned} M_0 = &- \frac{\kappa}{\kappa+1}\operatorname{Im}\{\sum_{j=1}^{p}(X_j - iY_j)z_j\} \\ &+ \frac{2}{\pi(\kappa+1)}\operatorname{Re}\{(\omega_1\bar{\omega}_2 - \bar{\omega}_1\omega_2)\sum_{j=1}^{p}(X_j + iY_j)\zeta(z_j)\} \\ &+ 2\operatorname{Re}\{\omega_2\psi_1 - \omega_1\psi_2\}. \end{aligned} \tag{6.54)'$$

We mention that, under the condition (6.55), $g(z)$ is determined up to a translation and so its addends g_1, g_2 are uniquely determined.

A relation among ψ_1, ψ_2, F_1 and F_2 may be obtained by (6.53) and (6.54)′. In fact, under the assumption (6.55), it is

$$\begin{aligned} \operatorname{Re}\{\omega_2\psi_1 - \omega_1\psi_2\} = &\frac{\kappa}{2(\kappa+1)}\operatorname{Im}\{\sum_{j=1}^{p}(X_j - iY_j)z_j\} \\ &- \frac{1}{\pi(\kappa+1)}\operatorname{Re}\{(\omega_1\bar{\omega}_2 - \bar{\omega}_1\omega_2)\sum_{j=1}^{p}(X_j + iY_j)\zeta(z_j)\} \\ &+ \operatorname{Im}\{\bar{\omega}_1F_2 - \bar{\omega}_2F_1\}. \end{aligned} \tag{6.56}$$

In particular, when the principal vector of external stresses on each L_j vanishes: $X_j + iY_j = 0$, (6.52) and (6.56) become respectively

$$\varphi_k = \frac{1}{\kappa+1}(iF_k + 2\mu g_k),\quad \psi_k = -\frac{1}{\kappa+1}(i\kappa\bar{F}_k - 2\mu\bar{g}_k),\quad k = 1, 2, \tag{6.57}$$

$$\operatorname{Re}\{\omega_2\psi_1 - \omega_1\psi_2\} = \operatorname{Im}\{\bar{\omega}_1F_2 - \bar{\omega}_2F_1\}, \tag{6.58}$$

and furthermore,

$$\begin{aligned} \operatorname{Re}\{\omega_2\psi_1 - \omega_1\psi_2\} &= \operatorname{Im}\{\bar{\omega}_1F_2 - \bar{\omega}_2F_1\} = \frac{2\mu}{\kappa+1}\operatorname{Re}\{\bar{\omega}_2g_1 - \bar{\omega}_1g_2\} \\ &= \frac{1}{2}M_0 = -\frac{1}{2}M_{L_0}. \end{aligned} \tag{6.59}$$

Moreover, from the above equatities, we may assure that

$$\operatorname{Re}\{\bar{\omega}_2\varphi_1 - \bar{\omega}_1\varphi_2\} = 0. \tag{6.60}$$

Let us then consider the case of cracks. As deduction as before, by noting that $H_j(z)$ is doubly-periodic, it is easily seen that (6.50) and (6.51) now become repectively

$$\begin{aligned} &\frac{1}{2\pi(\kappa+1)}\sum_{j=1}^{p}(X_j + iY_j)[\eta_k(a_j + b_j) - \kappa\bar{\eta}_k(\bar{a}_j + \bar{b}_j)] \\ &\qquad + \varphi_k + \bar{\psi}_k = iF_k,\quad k = 1, 2, \end{aligned} \tag{6.61}$$

$$\frac{\kappa}{\pi(\kappa+1)}\sum_{j=1}^{p}(X_j + iY_j)\operatorname{Re}\{\eta_k(a_j + b_j)\} + \kappa\varphi_k - \bar{\psi}_k = 2\mu g_k,\quad k = 1, 2, \tag{6.62}$$

and then (6.52) becomes

$$\left.\begin{aligned}\varphi_k &= -\frac{\eta_k}{2\pi(\kappa+1)}\sum_{j=1}^{p}(X_j+iY_j)(a_j+b_j)+\frac{1}{\kappa+1}(iF_k+2\mu g_k),\\ \psi_k &= \frac{\kappa\eta_k}{2\pi(\kappa+1)}\sum_{j=1}^{p}(X_j-iY_j)(a_j+b_j)-\frac{1}{\kappa+1}(i\kappa\bar{F}_k+2\mu\bar{g}_k),\end{aligned}\right\} k=1,\ 2. \tag{6.63}$$

Analogously, when the resultant of the principal vectors of the external stresses on the two sides of each crack L_j is zero: $X_j + iY_j = 0$, $j = 1, \cdots, p$, the equalities (6.57) – (6.61) remain unchanged.

Remark The foregoing discussions may be extended to the case where both holes and cracks occur in the elastic region without any difficulty.

2. Formulation of the fundamental problems

After the above discussions, we may give the general formulation of doubly-periodic fundamental problems. We would deal with the case of elastic region with holes. All the notations are as before.

For the *doubly-periodic first fundamental problem*, the external stresses $X_n(t) + iY_n(t)$, $t \in L$, is given. However, it is not sufficient for the uniqueness of solution of the elastic equilibrium. Besides, we should know, for instance, the principal vectors of stresses F_1 and F_2 on Γ_1 and Γ_2 respectively. Under these conditions, the uniqueness of the solution of the problem may be established.

In fact, if $X_n(t) + iY_n(t) = 0$ on L and $F_1 = F_2 = 0$, then it is easily seen that the "energy integral"

$$\begin{aligned} J &= \int_{L_0}(X_n u + Y_n v)\,ds + \int_{\Gamma_0}(X_n u + Y_n v)\,d\sigma \\ &= \mathrm{Re}\int_{\Gamma_0}(X_n + iY_n)g(\tau)\,d\sigma = 2\mathrm{Re}\{\bar{g}_1F_2 - \bar{g}_2F_1\} = 0, \end{aligned} \tag{6.64}$$

which follows that all the stresses in S^+ are equal to zero (Cf. Muskelishvili [1]). This is
equivalent to the uniqueness of the solution of the first fundamental problem. But we should note that, since in this case

$$M_{L_0} = \int_{L_0}(yX_n - xY_n)\,ds$$

is a given number and, by (6.53),

$$\mathrm{Im}\{\bar{\omega}_1F_2 - \bar{\omega}_2F_1\} = -\frac{1}{2}M_{L_0} \tag{6.65}$$

is also given, F_1, F_2 could not be arbitrarily given and have to fulfill (6.65).

The boundary condition for the first fundamental problem is

$$\varphi(t) + t\overline{\varphi'(t)} + \overline{\psi(t)} = f(t) + C_j(m, n),$$
$$t \in L_j(m, n) = \{t \mid t - \Omega_{mn} \in L_j\}, \tag{6.66}$$

where

$$f(t) = i\int_{t_j}^{t}[X_n(t) + iY_n(t)]ds, \ t_j, \ t \in L_j, \tag{6.67}$$

being doubly-periodically extended to L, and $C_j(m, n)$ are undetermined constants. Substituting (6.13) and (6.25) into (6.27), we get the boundary condition to be satisfied by
$\varphi_0(z)$ and $\psi_0(z)$:

$$\varphi_0(t) + m(t)\overline{\varphi_0'(t)} + \overline{\psi_0(t)} = f_0(t) + C_j(m, n), \ t \in L_j(m, n), \tag{6.68}$$

where $f_0(t)$ is already single-valued and doubly-periodic. This is equivalent to the case where each $X_j + iY_j = 0$, $j = 1, \cdots, p$, which would be assumed in the sequel. To solve this boundary value problem, in fact it is sufficient to solve

$$\varphi_0(t) + m(t)\overline{\varphi_0'(t)} + \overline{\psi_0(t)} = f_0(t) + C_j, \ t \in L_j, \ j = 1, \cdots, p, \tag{6.69}$$

where $C_j = C_j(0, 0)$, $j = 1, \cdots, p$, are undetermined, while the other $C_j(m, n)$'s may be naturally determined by the addends of $\varphi_0(z)$ and $\psi_0(z)$.

We assume the condition (6.55) is fulfilled. We have

$$\varphi_k + i\bar{\psi}_k = iF_k, \ k = 1, 2, \tag{6.70}$$

because of $X_j + iY_j = 0$, and, in (6.59),

$$M_{L_0} = \mathrm{Re}\{\int_{L_0} f_0(t)d\bar{t}\}.$$

Hence, once $\varphi_0(z)$ and $\psi_0(z)$ have been found out, it is easily proved that (6.57) is valid on considering M_0 by two different manners. And certainly (6.59) is also true.

To sum up, the doubly-periodic first fundamental problem may be reduced to: without loss of generality, under the assumption that the principal vector of the external stresses: $X_j + iY_j = 0$ on each L_j $(j = 1, \cdots, p)$, find two doubly quasi-periodic holomorphic functions $\varphi_0(z)$ and $\psi_0(z)$ satisfying the boundary condition (6.69) (where C_j, $j = 1, \cdots, p$, are undetermined constants), with given principal vectors F_1, F_2 of the external stresses on Γ_1, Γ_2 respectively with respect to S_0^+ fulfilling the condition

$$\mathrm{Im}(\bar{\omega}_1F_2 - \bar{\omega}_2F_1) = -\frac{1}{2}\mathrm{Re}\{\int_{L_0} f(t)d\bar{t}\}, \tag{6.72}$$

and, supplementarily, the addends of $\varphi_0(z)$ and $\psi_0(z)$ ought to fulfill (6.70). Here, we have assumed (6.55) is satisfied. Moreover, it is easily verfied, for the uniqueness of the solution $\varphi_0(z)$, $\psi_0(z)$ (not that of the stress distribution), we may require furthermore $\varphi_0(0) = \psi_0(0) = 0$. If $C_1 = 0$ is preassigned, then only one condition $\varphi_0(0) = 0$ or $\psi_0(0) = 0$ could be required.

Condition (6.60) and (6.58) may serve as check formulas for φ_k, ψ_k.

For the *doubly-periodic second fundamental problem*, given the displacement function
$g(t)$ on each $L_j(j = 1, \cdots, p)$ and its addends g_1, g_2, find the elastic equilibrium. In this case, $X_j + iY_j$, $j = 1, \cdots, p$, are undetermined. It may be transferred to solve the doubly quasi-periodic boundary value problem

$$\kappa\varphi_0(t) - m(t)\,\overline{\varphi_0'(t)} - \overline{\psi_0(t)} = g_0(t),\ t \in L_j,\ j = 1, \cdots, p, \tag{6.73}$$

where $g_0(t)$ contains terms involving $g(t)$ as well as undetermined constants $X_j + iY_j$, $j = 1, \cdots, p$, but there is no supplementary requirement similar to (6.72) for g_1, g_2 instead of F_1, F_2. For the uniqueness of $\varphi_0(z)$, $\psi_0(z)$, it is sufficient to require $\varphi_0(0) = 0$ or $\psi_0(0) = 0$, which is easily proved.

The existence of the solution of the first or the second fundamental problem may be realized in the method of solution, for instance, by reducing it to certain Fredholm integral equation as in the non-periodic case (Cf. Lu [6]), with kernel containing Weierstrass ζ- or $\wp$-function.

Similarly, we may also consider doubly-periodic fundamental problems of region with cracks, where $\varphi_0(z)$ and $\psi_0(z)$ are permitted to possess integrable singularities at the tips of the cracks. There is no difficulty in principle for solution of the problem which may be reduced to singular integral equation with Weierstrass ζ-function appeared in its kernel.

Bibliography

*(References with * or ** were published in Chinese or Russian respectively, somtimes with English abstract)*

Ahlfors L.V.

1. Complex Analysis, MacGraw-Hill, New York, 1966

Babuskha I. et als

1. Mathematische Elastizitätstheorie der Elbenen Probleme, Berlin, 1960

Baker C.T.H.

1. The Numerical Treatment of Integral Equations, Oxford, 1977

Bueckner H.F.

1. *Some stress singularities and their computation by means of integral equations*, in "Boundary Problems in Differential Equations" (ed. by R. E. Langer), 215 – 230, Madison, 1960

Cai Hai-tao

1*. *On periodic Riemann-Hilbert boundary value problems of half-plane*, J. of Central South College of Mineralogy and Metallurgy, 1979, No. 1, 105 – 112

2*. *Periodic contact problems in isotropic plane elasticity*, J. of Central South College of Mineralogy and Metallurgy, 1979, No. 1, 113 – 116

3*. *Periodic contact problems in plane elasticity*, Acta Math. Appl., **2** (1979), 131 – 195

4*. *On the first and the second periodic fundamental problems of infinite half-plane with anisotropic medium*, Acta Mech., **11**(1979), 240 – 247

5*. *Periodic crack problems in anisotropic plane elastic medium*, Acta Math. Sci., **2**(1982), 35 – 44

6. *A periodic arrary of cracks in an infinite anisotropic medium*, Engineering Fracture Mechanics, **46**(1993), 127 – 131

7. *The crack problem of two bounded half orthotropic planes materials*, Engineering Fracture Mech., **52**(1995), 895 – 900

8. *An application of complex analysis to periodic movable loading problems*, Complex Variables, **30**(1996), 145 – 151

Chibrikova L. I.

1**. *On Riemann boundary problem for automorphic functions*, J. Kazan Univ., **116**(1956), 59 - 109

Dmowsha R., Kostrov B. V.

1. *A shearing crack in a semi-space under plane strain condition*, Arch. Mech. Stas., **25**(1973), 421 - 440

Dundurs J., Tsai K. C., Keer L. M.

1. *Contact between elastic bodies with wave surfaces*, J. Elast., **3**(1973), 109 - 115

Erdogan F.

1. *Mixed boundary-value problems in mechanics*, Mechanics Today, **4** (1978), 1 - 87
2. *Fracture problems in composite materials*, J. Engng. Fract. Mech., **4** (1972), 311 - 340

Erdogan F., Gupta G. D.

1. *On the numerical solution of singular integral equations*, Quart. Appl. Math., **29**(1972), 525 - 534

Filschtisky L. A.

1**. *Doubly-periodic problem in theory of elasticity for isotropic media, weakened by a congruent set of arbitrary holes*, PMM, **36**(1972), 628 - 690

Gakhov F. D.

1**. Boundary Problems, Nauka, Moscow, 1977

Galin L. A.

1**. Contact Problems in Theory of Elasticity, Moscow, 1970

Gladwell G. M. L.

1. Contact Problems in the Classical Theory of Elasticity, Noordhoff, 1980

Gupta G. D., Erdogan F.

1. *The problem of edge cracks in an infinite strip*, J. Appl. Mech., **41** (1974), 1001 - 1006.

Hao Tian-hu

1*. *A solution in closed form of the anti-planar problem for the field of doubly-periodic cracks*, J. of Qinghua Univ., **19**(1979), No. 3, 11 - 18

Hao Tian-hu et als.

1*. *Anti-planar problems for the field of non-homogeneous doubly-periodic cracks*, Appl. Math. and Mech., **6**(1985), 191 - 196

Howland R.C.J.

1. *Stress in a plate containing an infinite row of holes*, Proc. Royal Soc. London, Ser. A, **148**(1935), 471 - 491.

Ioakimidis N.I.

1. *The numerical solution of crack problems in plane elasticity in the case of loading discontinuities*, **13**(1980), 709 - 716

Isida M.

1. *On some planar problems of an infinite plate containing an infinite row of circular holes*, Bull. JSME, **3**(1960), 259 - 265

Karihaloo B.L.

1. *Fracture of solids containing arrays of cracks*, Engng. Fract. Mech., **12** (1979), 49 - 77

Koiter W.T.

1. *An infinite row of collinear cracks in an infinite elastic sheet*, Ingenieur Archiv, **28**(1959), 168 - 172
2. *Stress distribution in an infinite elastic sheet with a doubly-periodic set of equal holes*, in "Boundary Problems in Differential Equations" (ed. by R.E. Langer), 191 - 213, Madison, 1960
3. *An infinite row of parallel cracks in an infinite sheet*, in "Problems of Continuous Mechanics" (Contributions in honor of the 70th birthday of N.I. Muskhelishvili), 246 - 259, Philadelphia, 1961

Krenk S.

1. *On the use of the interpolation polynomial for solutions of singular integral equations*, Ser. Mech. Appl., **32**(1975), 479 - 484
2. *On quadrature formulas for singular integral equations of the first and second kinds*, Ser. Mech. Appl., **33**(1975), 225 - 232
3. *Periodic contact and crack problems in plane elasticity*, Let. Appl. Engng. Sci., **4**(1976), 343 - 353

Kuznetsov E.A.

1. *Periodic fundamental mixed problem of elastic theory for a half-plane*, Prik. Mech., **12**(1976), No.9, 89 - 97

Lekhnitsky S.G.

1**. Anisotropic Plate, 1947, Moscow

Li Guo-ping

1*. Automorphic Functions and Minkowski Functions, Acad Press, 1979, Beijing

Li Xing

1*. *Doubly-periodic welding problems*, Thesis of Ningxia Univ., Ningchuan, 1988

Litvinchuk G.S.

1**. Singular Integral Equations and Boundary Problems with Shifts, Nauka, Moscow, 1980

Liu Shi-qiang

1*. *On periodic fundamental problems of an elastic strip*, J. of Math. (Wuhan), **4**(1984), 165 – 176

Lu Jian-ke (Lu Chien-ke)

1*. *Periodic Riemann boundary value problems and their applications to elasticity*, Acta Math. Sinica, **13**(1963), 343 – 388
2*. *Fundamental problems of infinite elastic plane with cracks*, J. of Wuhan Univ. (Sci. ed., Special Issure of Math.), 1963, No.2, 50 – 66
3*. *On fundamental problems of plane elasticity with periodic stresses*, Acta Mech. Sinica, **7**(1964), 316 – 327
4*. *Complex Airy functions in theory of doubly-periodic plane elasticity*, J. of Math. (Wuhan), **6**(1986), 319 – 330
5. Boundary Value Problems for Analytic Functions, World Scientific, Singapore, 1993
6. Complex Variable Methods in Plane Elasticity, World Scientific, Singapore, 1995

Ma Dao-wei

1*. *The periodic crack problems of compound materials*, Acta Math. Sci., **5**(1985), 31 – 41

Mandjavidze G.F.

1**. *On a singular integral equation with discontinuous coefficients and its application to theory of elasticity*, PMM, **15**(1951), 279 – 295

Mikhlin S.G.

1. Integral Equations and their Applications to Certain Problems in Mechanics, Mathematical Physics and Technology, Pergamon, New York, 1957

Morigashi S.

1. Theory of Plane Elasticity, Iwanami, Tokyo, 1961 (in Japanese)

Muskhelishvili N.I.

1. Some Basic Problems of the Mathematical Theory of Elasticity, Noordhoff, Groningen, 1953
2. Singular Integral Equations, Noordhoff, Leyden, 1977

Savin G.N.

1*. Stress Concentration around Holes, Sci. Press, Beijing, 1958

2**. *Stresses in elastic plane with an infinite row of equal cuts*, Dokl. USSR, **23**(1939), 515 – 519

Savruk M.P.

1**. *Doubly-periodic system of cracks with prolonged motion in elastic body*, Practical Mech., **11**(1975), 113 – 117

Sih G.C., Liebowitz H.

1 . *Mathematical theories of brittle fracture*, in "Fracture" (edited by H. Liebowitz) **2**(1968), 67 – 190, Academic Press

Tang Li-min

1*. *Stress analysis of elastic plane with several neighboring holes*, Sci. Records, **10**(1959), 366 – 375

Theocaris P.S., Stassinakis A.

1. *Complex stress intensity factors at tips of cracks along interfaces of dissimilar media*, Engng. Frac. Mech., **14**(1981), 361 – 372

Zheng Ke

1*. *Fundamental problems in doubly-periodic plane elasticity*, Acta Math. Sci., **8**(1988), 95 – 104

2*. *Fundamental problems in the elastic plane with doubly-periodic cracks*, Acta Math. Sci., **8**(1988), 321 – 326

3*. *Periodic crack problems for bonded half-planes of different media*, Acta Math. Engng., **10**(1993), 8 – 16

4. *On the mixed problems in an elastic plane with periodic cracks*, J. of Math. (Wuhan), **14**(1994), 33 – 40

5. *On the fundamental problems in an infinite anisotropic elastic plane with periodic cracks*, Proc. of the Sixth China-Japan Symposium on BEM, Nov. 1994, 19 – 24

Zhou Cheng-ti et als

1*. *Stress calculations on elastic plane with infinitely many circular holes arranged in a net* (I), J. of. Dalian Engng. College, 1960, No. 1, 99 – 125

Index

(The adjective "periodic" in each of the related terms is omitted)

www.ingramcontent.com/pod-product-compliance
Lightning Source LLC
Chambersburg PA
CBHW020214290326
41948CB00001B/40
* 9 7 8 9 0 5 6 9 9 2 4 2 2 *